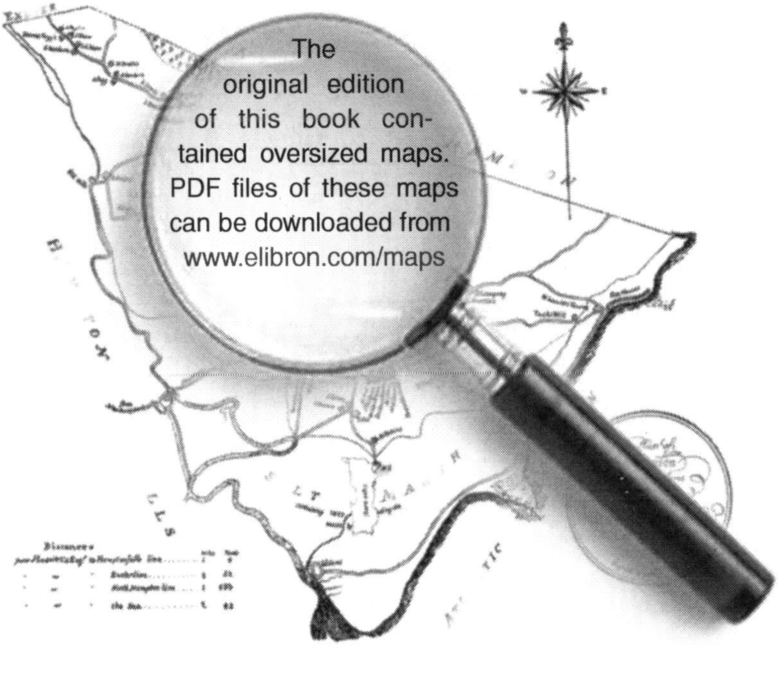
The
original edition
of this book con-
tained oversized maps.
PDF files of these maps
can be downloaded from
www.elibron.com/maps

ROBERT M'CORMICK

VOYAGES OF DISCOVERY
IN THE
ARCTIC AND ANTARCTIC SEAS
AND
ROUND THE WORLD

VOLUME II

Elibron Classics
www.elibron.com

Elibron Classics series.

ISBN 0-543-97047-7 (paperback)
ISBN 0-543-97046-9 (hardcover)

This Elibron Classics Replica Edition is an unabridged facsimile of the edition published in 1884 by Sampson Low, Marston, Searle, and Rivington, London.

VOYAGES OF DISCOVERY.

VOL. II.

LONDON:
PRINTED BY GILBERT AND RIVINGTON, LIMITED,
ST. JOHN'S SQUARE.

Vincent Brooks, Day & Son, Lith.

THE AUTHOR

ÆT 25.

From a portrait by Midshipman Skinner. Drawn on board of H.M.S. "Icarus."

FRONTISPIECE TO VOL. II.

VOYAGES OF DISCOVERY

IN THE

ARCTIC AND ANTARCTIC SEAS,

AND

ROUND THE WORLD:

BEING PERSONAL NARRATIVES OF ATTEMPTS TO REACH

THE NORTH AND SOUTH POLES;

AND OF AN

OPEN-BOAT EXPEDITION UP THE WELLINGTON CHANNEL IN SEARCH OF

SIR JOHN FRANKLIN

AND HER MAJESTY'S SHIPS "EREBUS" AND "TERROR,"

IN HER MAJESTY'S BOAT "FORLORN HOPE,"

UNDER THE COMMAND OF THE AUTHOR.

TO WHICH ARE ADDED AN

AUTOBIOGRAPHY, APPENDIX,

Portraits, Maps, and numerous Illustrations.

BY

DEPUTY INSPECTOR-GENERAL R. M'CORMICK, R.N., F.R.C.S.

Chief Medical Officer, Naturalist, and Geologist to the Expeditions.

IN TWO VOLUMES.

VOL. II.

London:

SAMPSON LOW, MARSTON, SEARLE, AND RIVINGTON,

CROWN BUILDINGS, 188, FLEET STREET.

1884.

CONTENTS.

Part III.

VOYAGE IN SEARCH OF SIR JOHN FRANKLIN AND THE CREWS OF H.M.S. "EREBUS" AND "TERROR."

CHAPTER IV.

CHAPTER V.

CHAPTER VI.

CHAPTER VII.

CHAPTER VIII.

CHAPTER IX.

CHAPTER X.

CHAPTER XI.

CHAPTER XII.

CHAPTER XIII.

CHAPTER XIV.

CHAPTER XV.

CHAPTER XVI.

CHAPTER XVII.

CHAPTER XVIII.

CHAPTER XIX.

APPENDIX.

LIST OF ILLUSTRATIONS AND MAPS.

PART III.

VOYAGE

TO

WELLINGTON CHANNEL

IN SEARCH OF

SIR JOHN FRANKLIN

AND THE CREWS

OF

HER MAJESTY'S SHIPS

"EREBUS" AND "TERROR."

1852-3.

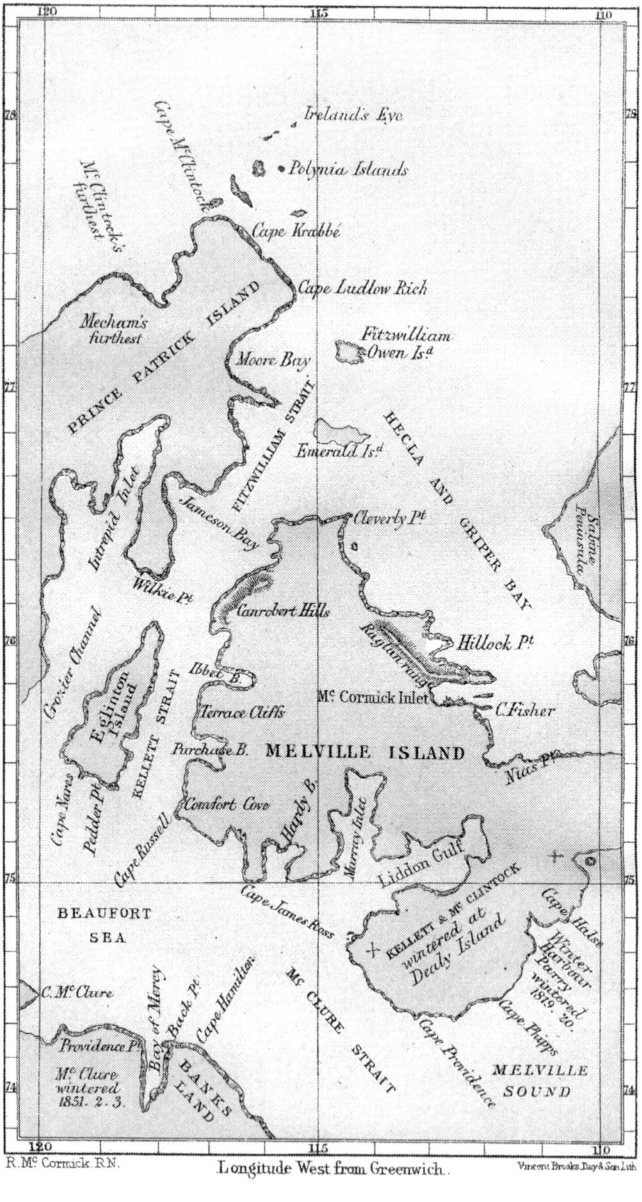

MAP OF MELVILLE ISLAND &c.

TO FACE TITLE PAGE
TO FRANKLIN SEARCH VOL.II.

Arctic Squadron at anchor, Whale-fish Islands. (*See page 7.*)

PERSONAL NARRATIVE

OF THE

VOYAGE

IN SEARCH OF

SIR JOHN FRANKLIN

AND THE CREWS

OF

H.M.S. "EREBUS" AND "TERROR."

CHAPTER I.

Departure from the Nore — Stromness — Greenland — Remarkable Felspar bands—Esquimaux—Rambles over the island—A hazardous descent—Departure for Disco.

Wednesday, *April* 21*st*, 1852.—At five a.m. the *North Star* was taken in tow by the *Lightning* steamer, and the rest of the Arctic squadron, *Assistance*, *Resolute*,

Intrepid, and *Pioneer*, by the steamers *Desperate* and *Basilisk*, and left Greenhithe to follow up the search for Franklin.

At nine a.m. we anchored off the little Nore. A lovely, bright, sunny morning. Captain Charles Hope, of H.M.S. *Monarch*, at Sheerness, came on board, and inspected the ship. Twenty years ago, when he was in command of the *Tyne*, of twenty-eight guns, I returned from Rio Janeiro to England as a passenger on board that ship, and, on my making myself known to him, he shook me most cordially by the hand, heartily wishing me every success in my present undertaking.

Thursday, 22nd.—At five a.m. weighed, and were taken in tow by the *African* steamer till two p.m., when we cast her off, and with a fair wind, and under a press of canvas, ran along the Suffolk and Norfolk coasts. Saw Alborough, Lowestoft, &c., and between five and six p.m. passed outside of Yarmouth Roads, at the back of the sands, having a distant view of Nelson's Monument, St. Nicholas Church, Gorlston Church, and the windmills on the Denes.

At eight p.m. I saw Winterton Light on the port quarter, and soon after the two lights of Hasborough. On quitting the deck to turn in, at ten p.m., I saw Cromer Light, and also the light placed on the Hasborough Sands, since I was last along this coast, now twenty-five years ago, in H.M.S. *Hecla*, in Parry's attempt to reach the North Pole. During the night, as we entered upon the wider expanse of waters of the North Sea, out of sight of land, we encountered a short, tumbling sea, in which the ship rolled heavily, in a fresh gale of wind, under double-reefed topsails. Lost sight of the rest of the squadron astern.

Saturday, 24th.—Off Buchaness Head, and took a sketch of it. On the following day, Sunday, 25th, at nine a.m., anchored off Stromness.

Monday, 26th.—At eleven a.m. I landed at Stromness, and on entering the "Commercial Inn," I met Lieutenant Pim, of the *Resolute*, there, who introduced me to Dr. Hamilton, a resident in the place, who was married to a sister of Dr. Rae, the Arctic traveller, and had himself been formerly in the navy. I took a sketch of the town and harbour of Stromness.

Wednesday, 28th.—Sailed at two p.m., passing the bold headlands of Hoy and the Black Crag. Took a sketch of the Old Man of Hoy.

Wednesday, May 5th.—Blowing a hard gale of wind, in which the ship rolled heavily. One of these seas struck her on the port forechains, stove in three of the ports, part of the hammock netting, the gangway door, and the bows of the boat in the waist, at the same time the starboard quarter-boat filled, when the hook giving way, she, with all her gear, was washed away. This sea, after flooding the main-deck everywhere below, and putting out the galley fire, carried overboard with it one of our hapless pigs.

Saturday, 8th.—The steamer, after having supplied us with a cutter to replace the boat we lost in the recent gale, parted company with the squadron at noon to-day.

Monday, 10th.—I saw the first mallemuck (*Procellaria glacialis*), and in the evening a flock were flying round the ship. A copy of Gilley's "Shipwrecks of the Royal Navy" happening to be in the ship's library, I, for the first time in my life, read the account of the melancholy loss of H.M.S. ships *Defence* and *St. George* on that fatal reef off the coast of Jutland, when returning with the Baltic Fleet, on Christmas Eve, in the year 1811, when my poor father perished in the first-named ship.

Thursday, May 20th.—At 5.30 a.m. I went on deck to see the first piece of ice, just visible on the margin of the horizon, in the form of three small hummocks, some two leagues distant, we having just entered within Davis's

Straits. Saw a snow bunting, two little auk, a mallemuck or two, and a small flock of kittiwakes.

Friday, 21st.—At eleven a.m. saw the land in the vicinity of Cape Comfort, four or five leagues distant, on starboard beam; but unluckily the weather became so thick and misty, with a driving sleet, that its snow-capped summit was too indistinctly seen to get a sketch of the cape.

Monday, 24th.—At 4.30 a.m. I went on deck to have a look at the land. The bleak, barren-looking coast of Greenland appeared on the starboard beam at a distance of some four or five leagues, extending along the horizon in a line of black and white cliffs, piebald with the mantling of snow, a white haze concealing the base. One remarkable and lofty peak appeared to be intersected from the summit to the base by a large trap dyke. I took several sketches whilst beating up. Lat. at noon 64° 27′, long. 52° 31.′

Saturday, 29th.—A thick, foggy day, with snow falling and light winds, standing in for the Whale-fish Islands. As we approached the island at the back of which the settlement is situated, the dense curtain of mist lifted for a short interval, enabling me to get a complete sketch of the land on either side. High hills on the port, and low-lying islands extending along the horizon on the starboard side, interspersed here and there with a berg or small stream of ice. The solitude was enlivened by the shrill and animated notes of a flock of tern, which, with two or three flocks of the eider-duck, and a few looms swimming about, gave welcome variety. I had scarcely finished my sketch when the horizon all round became suddenly overcast with a continuous fall of snow, until after we had anchored in the harbour. When we were some three or four miles from the anchorage nine canoes came off and paddled around us, one only was of a white colour from the effects of age, the others were dark.

The Esquimaux in their yellow-seamed, dark jackets, and round-topped caps, with a sort of French peak, and harpoon lines coiled on a tripod in front of them, a spear on either side, and a dark-skin bag inflated for a buoy, in the stem at the back of them, altogether presented a picturesque scene. We anchored in a perfectly land-locked harbour, the granitic cliffs all round scarcely attaining an altitude of 100 feet, clothed with a brownish vegetation between the patches of snow. As we hauled round the point of the island, on which a kind of flagstaff of the whale's jaw-bone was erected, several groups of Esquimaux had assembled, running from rock to rock to get the best view of the ships coming in. Indeed the Arctic squadron of five vessels must have presented a very imposing spectacle to these fur-clad children of the floe, some of whom came aboard.

Sunday, 30th.—I went on board the *Resolute* to see old Captain Kellett, who had been so ill as to have been confined to his cabin the whole of the way out. I found him sitting up in his cabin, with several officers around him. He was looking much reduced in flesh and strength, but in excellent spirits, joking with those around him, and telling me that I was to be knighted on my return.

Monday, 31st.—Fine day with a temperature of 38° Fahr. I landed on the island on which the Danish settlement is located, accompanied by my young friend, Frank Toms, the assistant-surgeon of the *North Star.* On the beach several Esquimaux canoes were lying near a small, rude storehouse, called by the Dane an ale-house. This Dane, by name Fritz Emiel Torgenfead, and styled the governor, lives in a small hut a little higher up the hill, surrounded by the putrid carcases of seals and other filth. A group of Esquimaux women and about half a score of children were squatting about outside, with several fine dogs of mixed Danish and Esquimaux breed. On entering the hut we passed through two small anterooms, containing a sledge and harness for the dogs, a few fresh dog-skins, and the wooden framework of a small canoe not yet covered with sealskin, and intended for the use of the child of the Dane, whom we found in the innermost apartment, seated near a small table close to the only window, formed of a dozen panes of glass, and employed reading a Dutch prayer-book alone. He told me that he was married to an Esquimaux woman, which would be a barrier to his return to his native land, "Huskey" women not being permitted to be naturalized in Denmark. He had resided here two years. The settlement, he informed us, contained altogether about 100 natives, including the children. The winter is the great fishing season, little being done in that way in the summer months. He had two narwhale's horns in his possession, which we bought of him, each of us one.

The furniture of this apartment consisted of a broad bench of boards, loosely covered with a few sealskins, occupying one side of the room as a bed-place, and on the other side a small deal table and a chair, with an old guitar suspended from the wall, which was papered with Danish newspapers, and the solitary Dutch prayer-book constituted the library of this poor, simple-minded recluse. He very civilly accompanied us down to the beach to show us his canoe, the doors of the hut closing after us with a weight as we made our egress. The canoe, however, had gone off to the ships, and we did not see it. I thought of purchasing it, but he asked too much—twenty dollars Danish money, upwards of 2*l.* sterling. After I had taken a sketch of the harbour with the ships in the distance, we continued our excursion over to the west side of the island, which is here very narrow. Four or five of the Esquimaux dogs voluntarily accompanied us round the island, presumably as guides, for others tied up set up such a tremendous howl in concert on our departure, as if they too had an equal claim to accompany us.

The whole island is of granitic formation, and with the broad bands of salmon-coloured felspar, and here and there veins of pure white quartz intersecting the face of the rocks, presenting sections of a grey, salmon, and white colour, embossed in the hollows with the scanty vegetation of mosses, lichens, and the dwarf willow, creeping along the barren soil amongst the smooth-looking masses of water-worn granite, have a very picturesque effect, to which the deep fissures or gorges in the rocks where the surf breaks in, add another and additional wildness to the scene. We only met with three or four snow-buntings on the island, but saw a few tern gracefully skimming the surface of the sea, which was studded by several icebergs. The lively, musical note of the pretty agile little tern relieved the sombre solitude around. Having taken a diagram of this remarkable veining of the rocky coast and collected a few specimens, we returned to the landing-place, where we again met the Dane, and gave him an invitation to come on board and see the ship. Our old quarter-master came for us in the dingy, and we got on board to dinner at three p.m.

After dinner, accompanied by Alston, the mate, pulled in the dingy round the island to the eastward for about two miles, landing up a creek, but a fog coming on compelled us to return on board at seven p.m.

Tuesday, June 1st.—Very fine day, with strong breezes. At 9.30 a.m. I landed opposite the ships on the north island, ascended the ridge of hills above, and descended on the opposite side into a valley or ravine running east and west; the cleavage planes of the plutonic rocks dipping to the east at an angle of 20°, which appears to be the general bearing and direction of the ridges and valleys. The hollows were many of them deeply filled with snow-drift. Ascending another hill and collecting some rocks, and filling my botanical case with plants, I left them on the summit after taking the bearings of the surrounding islands, and then descended into another ravine, and eventually reached the summit of the highest hill on the island, some 200 feet in altitude.

I now retraced my steps from this the centre hill of the island, and recrossing the ravine to the summit of the first hill, where I had left my case of plants and rocks, I freighted myself with them, and returned to the harbour, getting on board at two p.m. After dinner, accompanied by Toms, I took the dingy and pulled as far as the point at the west end of the south island, and landing, ascended the high ridge of rocks above the anchorage of the ships, and down to the valley on the opposite side, from which I returned by a steep, narrow gorge in the almost perpendicular rocks, down to the water's edge, opposite the ships, where Toms was in readiness with the boat to take me on board, just in time for tea, at six p.m. A party of the Esquimaux, men and women, came alongside this evening in a large boat, having a little girl with them, dressed in the same costume as her seniors. I took a sketch of the upper and lower ends of the harbour before I turned in at midnight.

Friday, 4th.—A very fine day. I landed on the small island west of the Esquimaux Settlement Island, and separated from it by a narrow strait. A number of tern were hovering around me, so vociferous that they must have had eggs in the vicinity, although, after a very diligent search, I could not discover any, after walking over the whole island. I collected a few plants and rocks, and took a sketch of the anchorage with the ships, returning on board in the dingy at 1.30 p.m. After tea, and just as I was about leaving the ship in the dingy with Toms and Alston, a young Esquimaux and his father came alongside, and their canoe was the most symmetrical in form, and newer in appearance than any I had hitherto seen, but they would not sell it. Pulled for the second island from the settlement, and having landed there, I ascended the highest hill geologizing, whilst my companions amused themselves in shooting tern. We next pulled across to the opposite shore, and I landed on the farthest point, through a heavy swell and at the expense of a ducking in the surf. I ascended to the summit of this portion of the North Island with ease, and crossed over to the precipitous cliffs beneath which the boat with my three companions was waiting to pick me up on my return.

I had before me a most difficult and hazardous descent to accomplish down the very face of a cliff some 200 feet in height, and almost as perpendicular as a wall, broken only by vertical ledges of rock intersected by deep chasms and fissures, made slippery with ice or running streams of water, with hollows and cracks filled with snow-wreaths. At one moment I was crawling on all fours, over mural ledges of rock, and at another leaping from crag to crag, or sinking knee-deep in some hollow of snow. I at last, after repeated efforts, succeeded in reaching the beach below in safety, upon which I should have been very summarily hurled instantly, had I made one false step in my perilous progress; nevertheless I managed to snatch up a few specimens of mosses, plants, and rocks in my downward course, as mementoes of this little adventure, and to complete the natural history of these isles. My companions in the boat loudly cheered me on, when they caught sight of me in the most perilous part of my descent, on the precipitous ledges just below the summit.

Saturday, 5th.—The squadron weighed at seven a.m., and with bright, sunny weather, stood out of the harbour from the Whale-fish Islands, by the North Channel, in the direction of Disco—the bold, high, and remarkable black cliffs of which appeared to the eastward, about five leagues distant, and off which we arrived in the evening, but did not let go the anchor until eleven p.m.

Several large and remarkable-looking bergs were aground off Disco, one having an archway right through it; another with deep caverns like church porches, at the entrance; one resembling a cottage with a garden wall, another in the shape of a sphinx. The sea here is open, and free from floe ice; two eider-ducks passed us. On the point entering the harbour on the starboard hand, the Danish flag, swallow-tailed in shape, having a red ground with a white cross, was flying over a sort of tent, cask-shaped.

Sunday, 6th.—Foggy weather; after breakfast got up the anchor again, to shift our berth farther up the harbour; and, what has been no very unusual thing with the *North Star*, got on shore on the rocks, on the weather side, within thirty yards of the beach, when warping up, eleven a.m., having a rock nearly awash at high water, a few yards only under the bows. The ship heeled over to a considerable angle to port, and did not get off till the next high tide, at nine p.m. I saw a raven to-day for the first time during the voyage.

Back of Disco Mountain, and place of ascent. (*See page* 19.)

Glaciated Ancient Crater, Disco Strait. (*See page* 24.)

CHAPTER II.

Disco—A curious breakfast—The native waitress—The governor—His house—My dogs Erebus and Terror—An Esquimaux ball—Ascent of Disco mountain—A dangerous enterprise—I reach the summit—Specimens—The squadron again under weigh.

Monday, June 7th.—A very fine day; after breakfast I went on shore to call on the governor, accompanied by Toms. We were very kindly received, and a second breakfast was ordered for us, which possessed the charm of novelty from the various dainties of native produce placed before us. Eider-duck dressed in a form much resembling a beef-olive, and having much the same flavour. Another dish consisted of the inner portion of the skin of the whale, preserved in vinegar, and presenting a curious piebald appearance, the strips or ribbons into which it was cut being half black and half white, but somewhat insipid to our taste. A third dish was a pudding, made of cod-fish and currants, having an indescribable taste. With all these delicacies we had each a cup of coffee, Danish biscuits, and brown bread, and a cordial with some ginger gin completed this unique repast. Not less novel and interesting to us was the appearance of the young Esquimaux girl who placed these good things before us.

She was a fresh, healthy, good-looking girl of sixteen, with a skin as fair and cheeks of as ruddy a colour as any English country girl. Her hair, of a light-brown colour, was twisted up into a top-knot at the crown of the head, and bound round by a yellow ribbon. Her dress, a pink-checked, linen jacket, with the collar and cuffs trimmed with fur; seal-skin breeches, and white skin boots meeting these at the knee, having a band of red colour round the calf, and surmounted by white linen legging, which, with small rings in her ears, completed her toilet.

The governor appears to be a man of about fifty years of age, a tall, stout, manly-looking Dane, with the usual light hair of his country, wearing a small black silk skull-cap on the crown of his head, and dressed in a purple-check, linen jacket, with fur collar and cuffs, and seal-skin trousers.

The entrance to the governor's apartment is through a long winding dark passage from the front door, the floor of the room boarded, and having a rug under the table, which was placed next to a green-baize-covered sofa, and on the blue-coloured wall above it three engravings of landscapes in frames were suspended, with a looking-glass and a mirror between the two windows. These windows were small, formed of a dozen panes of glass, and overlooking the harbour and the steep, lofty hills on the opposite side. On the right of the sofa stood a buffet containing cordials and liqueurs, and between this and the door the tall, massive, iron stove, in the shape of a pillar, in which the fire is not seen had there been one, being quite enclosed. On the opposite side of the room stood a chest of drawers, surmounted by a coloured view of the settle-

Edwin Wilson, del. Vincent Brooks, Day & Son, Lith.

Erebus Terror.

ESQUIMAUX DOGS.

PAGE 14—VOL. II.

ment of Upernivik, representing a Moravian chapel, and some four houses backed by an angular-shaped hill, resembling a pyramid. Of this settlement, the governor told me, he had been in charge for some seven years previous to his coming here. Beneath this picture was suspended a daguerreotype of a son of his, now in Copenhagen for his education.

On either side of this were engravings from Schiller; on the left of the sofa stood a writing-desk, and half a dozen walnut-backed chairs, with red-flowered chintz bottoms. A door from this opened into the bedroom at the back, in which stood a small bedstead without curtains, and numerous papers scattered about a table at one corner. We did not see his wife, but his little boy of eight or nine years of age made his appearance at breakfast.

I afterwards accompanied the governor to the store-house, containing whale-lines, seal-skin jackets, biscuits, corn-beef, and pork, and he also showed me over the new house building for the inspector-general of this district. It contains eight rooms on the ground floor, and room for as many more on the upper one, each room being lighted by glazed windows. A piece of ground is enclosed in at the back for a garden.

I next visited the blacksmith's shop, to purchase three dogs for the use of the expedition. I also bought two pups for myself, about four months old, brother and sister, (the young Esquimaux dog being the colour of the polar bear, a creamy white, and the bitch a prettily-marked black and white,) for two Danish dollars, or 5*s.* sterling. I named them at once after my old ships, of which we are now in search, the dog, Erebus, and the bitch, Terror.

Afterwards I visited the Esquimaux hut, where I found the natives all assembled for a dance, the young girls waltzing with some of the officers of the expedition, in which my young friend Toms was taking a very prominent part. The governor's secretary was also one of the number. The old Esquimaux woman, who appeared as the duenna, was seated on the broad bench which serves the purpose of a bed at night and a couch or sofa by day. Alongside of her I took up my position, to witness the proceedings of a ball taking place under circumstances of so novel a nature. An engineer belonging to one of the tenders was playing the fiddle, to which the girls kept excellent time, and seemed to enjoy the whole very much. The governor's attendant, whom we saw at breakfast, by name Marie, with her sister Sophie, also a very pretty girl, a year or two older than her sister, and evidently the belles of the settlement, were eagerly sought for in the dance as partners.

From this lively scene I proceeded to one of the huts, to see an Esquimaux woman, the wife of one of the Danes, who was confined to her bed by an injury to the knee, and after going on board for my lancet-case, I opened a large abscess, which had been a source of much suffering to the poor creature. Afterwards I ascended the hill above the flagstaff, and called on the governor. Returned on board to tea. In the evening, the governor, accompanied by his secretary, came on board and remained until near midnight, and the whole of the Esquimaux women, who had been to dances got up on board the *Resolute* and *Assistance*, came on board of us, and finished their day's amusement with us. Waltzes and reels were kept up on the quarter-deck, in which both officers and men joined, with great spirit, till midnight, the gun-room steward playing the bugle.

The girls drank large quantities of water whilst dancing, and sang well, having good voices, with an admirable ear for keeping time. Wine, gin, and rum were handed to those who drank them, the men and the old women. The governor and his secretary played tunes on the violin, and sang well in our mess-room, where they partook of wine and smoked cigars. The whole party left in their canoes at midnight.

Tuesday, 8th.—I went on shore and called on the governor and on my Esquimaux patient, and afterwards, accompanied my young friend Toms to the Esquimaux hut, where the dance had taken place yesterday. We found the girls all very busy at their work, making leather watch-cases and other trinkets, and one old woman was employed in dressing the hair of a young girl just entering her teens. When combed down over her shoulders, her hair, which was of different shades of colour, was so long as to reach down to her heels. The old duenna at the dance was at work in her customary corner.

After some persuasion, I succeeded in getting Sophie to part with a pair of her best white boots. For these I gave her five shillings. Another pair, of a yellow colour, I bought of another woman for four shillings. The interior of the hut was rendered exceeding warm by an iron fireplace closed up. A gay-looking Dutch clock, surmounted by a group of red roses, ornamented the wall near the door, and some plates of ladies' fashions the opposite end of the apartment. The room was lighted by glazed windows.

I went to the Danish blacksmith's and paid for my two Esquimaux dogs, directing that they should be sent on board this afternoon. I gave the governor's secretary a monkey-jacket in exchange for a pair of Esquimaux fur boots. He kindly brought them himself down to the boat with me, and on the way I gave him a new silk kerchief—he having been much struck with the pattern on it—as a parting present.

Tuesday, June 8th.—At four p.m. I landed in the dingy at the top of the harbour, to ascend to the summit of the remarkable mountain opposite to the settlement, to ascertain its correct height above the level of the sea.

Our old quarter-master, Page, returned on board with the dingy, leaving me to accomplish the task I had imposed upon myself, alone and companionless in the mighty solitudes of these vast surroundings of mountains and ravines, filled by the countless ages of glacier ice and snows.

Having taken the height of my aneroid barometer on landing, which was the same as when I left the ship, 29° 32′, and the thermometer at 56° Fahr., I struck across the low, swampy ground, and rocky ridges, from the surveying mark, in the direction of the right spur of the mountain, passing over one rather steep ridge before I reached it at 5.15 p.m., when the aneroid indicated 28° 32′, the thermometer being 52°. Here I flushed a ptarmigan in its summer plumage, a mottled brown, and shot it, picking up a few plants on my way. I now fairly began the ascent of the right spur of the mountain, first over a steep slope of sandy *débris*, studded with loose fragments of rock. On gaining the summit of this ridge the ascent increased in steepness, and narrowed in breadth, and when I had attained to near the base of the craggy part of the upper portion of the mountain the ridge became so very sharp-edged, and the gusts of wind so violent that I was compelled to advance on hands and knees astride it, and on reaching the overhanging craggy precipice, found all further attempts quite impracticable, and out of the question in this direction.

In addition to the steepness, the fragments of rock were so loosely embedded in the *débris*, and so easily displaced by the weight of the body, which frequently had to be entrusted to their stability, that they fell crumbling into pieces on the least pressure, rolling down the precipices into the ravines with fearful velocity. On this spot, arresting as it did any further advance, I found a staff fixed in a small cairn of stones, most probably by some previous adventurer, who, to mark his failure, placed

R. McCormick, R.N., del.

Vincent Brooks, Day & Son, Lith.

Disco Mountain, above 2000 feet high, ascended by a new route from the back of the mountain.

PAGE 18—VOL. II.

it here; and I confess I was myself all but disposed to give it up too; but at six p.m., after taking the height of my aneroid, which indicated 27° 67′, and the thermometer 48°, I resolved that the last two hours of incredible labour and exertion should not be wholly thrown away, and I came to the determination at once to attempt to get round the back of the mountain on the inland side, in short to outflank it. Here only a narrow ledge was left for me to follow immediately beneath the overhanging, craggy summit, and the steep slope of the *débris* beneath. In this attempt, after encountering no small difficulties, at the momentary risk of some treacherous fragment giving way under my feet, and hurling me to the bottom of the precipice into the ravine below, a depth of nearly 2000 feet, I succeeded at last in reaching an angular projecting fragment of rock, about the centre of the back of the mountain.

This perpendicular pinnacle of projecting rock, extending between me and the summit, some twenty to thirty feet above my head, barred the narrow ledge on which I stood, and had been progressing with barely sufficient foothold to maintain my balance against the furious gusts of wind that directed their full force on this side of the mountain.

The only alternative now left me was either to retrace my steps after nearly four hours of such toil as I had already undergone, or risk the surmounting this pinnacle of cracked and decaying rock, dilapidated after ages of exposure to the climate in this region of everlasting frost and snow, till its crumbling condition presented a very frail hold indeed to trust to in scrambling up it. Having first thrown my old and favourite double-barrelled gun, with an effort, upon the secure flat surface of the summit above, I followed it with a spring, giving the tottering mass a kick as I did so, to aid the effort of propulsion to a secure footing on the very top of the mountain. It fortunately bore my weight, not a very great one, and I had the inexpressible gratification to find myself standing safe on its table-like top at eight p.m., after expending four hours in the attempt.

At 7.15 p.m., when immediately beneath the craggy precipice, the aneroid indicated 27° 12′, thermometer 46°, and now at the summit, when near the front edge overlooking the harbour, it fell to 26° 92′, and thermometer to 44°. At nine p.m., having crossed the top of a watercourse to a somewhat higher ridge, running backwards, I again took it, when it sunk to 26° 85′ thermometer still at 44°. On my subsequent return to the landing-place, where the boat landed me, I again took it, when it had risen to 29° 18′.

I remained for an hour and a quarter on the summit, but the greater height of the inland mountains entirely shut out all view of the Greenland coast, the prospect being bounded by hill-tops and deep ravines, filled with snow in every direction, a scene of glaciation. And well might Hayes call it the "land of desolation."

At the front of the mountain, immediately beneath me, lay the harbour of Lievely, dotted on the surface with the five vessels of our squadron, and a little Danish sloop. Beyond appeared the settlement on the small peninsula or rather island, forming the outer part of the land-locked harbour, and beyond this Baffin's Bay, with here and there a stream of ice, or a few scattered bergs, the Whale-fish Islands bounding the horizon to the westward. The long strip of land on which the settlement stands is deeply intersected by bights, with several small lakes or pools of water in its centre, one much larger than the rest. At two points of the western side I saw a hut or two of the Esquimaux. A few small islets skirt the margin outside.

The summit of this mountain is flat, rising into a ridge at the back, covered with snow, and separated

R. McCormick, R.N., del. Vincent Brooks, Day & Son, Lith.

The Danish Settlement of Lievely, Disco, North Greenland.

PAGE 20—VOL. II.

by a boggy indentation, with a watercourse. A few mosses, lichens and grasses, were sparingly scattered over the surface. The whole mountain is composed of plutonic rocks, greenstone, &c., intersected near the summit, by dark-reddish, horizontal bands, three or four of these bands containing a claystone rock, and occurring at regular intervals on the upper acclivity of the mountain, which may be divided into three divisions—the upper third, the craggy rock; the middle portion, loose, sandy *débris;* and the lower or base, rocky ledges and ridges. I saw the track of the footsteps in the snow, near the highest part, of Lieutenant Pim and his party, who had ascended on the previous day.

After gathering a few mosses and lichens, with a few rock specimens from the summit, I began the descent at 9.15 p.m., by the watercourse down the southern extremity, or opposite side of the mountain to the one at which I made the ascent, and found no difficulty; the centre of the watercourse was now filled with snow, and I kept for the most part of the way along the edge of the rocks, on the right side in my descent. On my reaching the bottom, a drizzling rain came on and continued until I got on board. Between the base of the mountain and the harbour, a distance of some two miles, the surface consists of rocky ledges, swampy marsh, and ridges of loose, sandy *débris*, studded here and there with fragments of rock. Here I collected a few more plants, and reached the rocks opposite to the ship, exactly at midnight, and our old quarter-master, who landed me, now came for me again in the dingy, and I got on board at 12.30, well freighted with rocks, plants. and the ptarmigan.

Found Lieutenant Pim with Toms in the cabin of the latter, anxiously awaiting my return. Pim told me he had found the ascent one of the most difficult and dangerous of any he had ever before attempted, and that himself and his party were five hours and a half in reaching the summit by the course which I descended it, comparatively an easy one. He, knowing that I was alone in my undertaking, was in some apprehensions for my safety. What would he have been had he been aware of the course I had taken? For it afterwards appeared that no one before had ever reached the summit by the course I took from the inland side of the mountain; the Danes have hitherto considered it quite impracticable. I accomplished it by this "new" route in the space of four hours from the landing-place, and I certainly was not more than two hours in the descent on the south side, by the track along the watercourse.

Thursday, 10th.—The Arctic squadron got under weigh this morning at breakfast-time. We were the first out of the harbour, but not a Dane or an Esquimaux appeared in the settlement to see us off. The weather was overcast and gloomy. The ships stood to the southward and eastward, along the west side of Disco, steering for the Waygatz Strait.

Friday, 11th.—Running along the coast of Disco, with light winds and calms. In the evening sporting parties were sent on shore in search of game, in which I joined, and had a pull of about two miles to the shore, landing on a fine sandy beach strewed with algæ, a few small pebbles, and some driftwood. Saw a dotterel. I made an excursion for about two miles inland, following a watercourse, but did not meet with a single ptarmigan, and only shot a male and female Lapland finch. The base of the mountains was distant about three miles over a level spongy bog, intersected by water-courses.

Whilst on shore this evening, one of our boat's crew came up to me and volunteered to be one of my own boat's crew, in my purposed expedition in search of Franklin. His name is Frost, a very appropriate one for the service we are going upon.. He is an active, hardy-looking young fellow, brought up in the coal and Baltic trade, and has already been in Hammerfest in Lapland. He is a native of Durham.

The aspect of the hills after leaving Lievely wears quite a different appearance. The whole line of the beach presents a smooth, white sand, and from this the rocks rise of a greyish-yellow colour, a dark greenstone range forming the background to the whole. Observed a large lake near the beach. Many of the flats and valleys presented a green appearance, with mosses and lichens. The island of Disco slopes gradually towards the Waygatz Straits, forming a very low and long narrow point, extending as far as the eye can reach beyond; and above which appears the mountain-range of Greenland itself.

My own observations of the mountain I ascended the other day, on being worked out, gave the height of 2100 feet, a very close approximation, I believe, to the altitude assigned to it by the Danes.

Saturday, 12th.—Within the Waygatz Strait, beating up against a very strong breeze, cold and biting to the fingers' ends. High land on both sides, with some remarkable peaks—the same greyish-yellow or pale-brown rocks, with the white, sandy beach—appear on this side of Disco as on the other; and the remarkable-looking, peaked, black mountain in the centre also makes much the same on this side, on rounding the first point. The strait really forms a curve or bay, but the lower half of the land was completely hidden in a stratum of mist very dense, and extending to the furthest extremity of the strait within sight. On the opposite side, the view was terminated by an abrupt black cliff. On the right or east side of the strait is a singular-looking, round-topped hill which, with some others, forms a circular cavity environed by enormous glaciers, and resembling a vast crater, probably some 4000 feet above the level of the sea, and the highest land in this neighbourhood. I took sketches of both sides of the strait. At one p.m. a signal was made to go about, and give up this dead beat to windward, against which, with such a strong breeze blowing, the ships made no headway, also orders to rendezvous at Upernivik.

Saturday, 19th.—After standing off and on, tacking about all night mid rocks, bergs, and shoals, in a most intricate navigation, with a heavy gale blowing in fitful violent gusts of wind, coming from off the high land of Sanderson's Hope, under the shelter of which we anchored at two p.m. in twelve or fourteen fathoms, on a rocky, bad-holding ground, with broken water. Not half a cable's length astern of us the sea is breaking over a sunken rock, right in our line of drift, but fortunately for us the ship did not drift, although in the height of the gale a berg got under weigh, and went out to sea with the hawsers we had made fast to it.

We got down the "crow's nest," and struck topgallant masts, pointing the yards to the wind. We had two anchors down, and had to wear out cable. In the latter part of the evening the furious gusts of wind were accompanied by a thick, drifting snow, almost blinding, and bitterly cold. Our situation was a very precarious one, on a wild and dangerous coast, surrounded by sunken rocks and a group of islands. Saw a Glaucus gull or two, and several ducks, looms, and dovekies.

Upernivik, Northernmost Danish Settlement of Greenland.

CHAPTER III.

Upernivik—The church—Governor's house—Barter with the natives—In Melville Bay—The *Resolute* nipped by bergs—A cold bath—We board the wreck *Regalia*—A whaling fleet—Grand march of the fishermen—Driving the dog-sledges—Visiting the whalers—A bear-hunt—On the march.

Sunday, June 20th.—I remained on deck all last night. The settlement of Upernivik consists of four or five houses, or stores, with a small white-fronted chapel on the sloping point of a hill near the beach. During the first watch eight Esquimaux canoes paddled out of the harbour to meet the ships. The rest of the squadron parted from the bergs they had made fast to, and drifted out to sea in the thick, hazy fog, and were some time in regaining their position. At six p.m. a letter-bag was made up, by which I sent letters home.

We have had a boisterous week's passage from Disco, the most disagreeable weather since we entered Baffin's Bay, mist and sleet, cold, raw, gloomy, and overcast, with streams of ice, and numerous icebergs of all shapes and dimensions; at times upwards of a hundred in sight at once. The coast of Greenland along which we passed presented generally a black aspect, especially Lawson Island, which is steep and bold, its black precipitous side intersected at intervals, close and regular, by horizontal reddish-brown lines. Took sketches of these, with Storoe Island, the Black Head, Cape Svartekuk, or the Black Hook. When beating up for Upernivik, Sanderson's Hope conspicuous.

After breakfast, accompanied by Toms, I landed from the dingy at the settlement. Met the Danish Governor, surrounded by a whole troop of Esquimaux dogs. He asked us into his house, furnished in a very inferior style to the Governor of Disco's. But here, as there, the violin was a conspicuous object, hanging against the wall with a few small engravings in frames. This governor appears to be from the same class as the one at the Whale Fish Islands, a mechanic—a cooper, I believe. He accompanied us into the church, where it appears some eighteen natives attend divine service. The pulpit, situated by the wall on the right of the entrance, had on either side a row of six benches, or forms, also on the left a stove, and at the further end the communion-table, of pretty-looking wood, rather handsomely gilded, and covered with a very white cloth, bearing on it two large candlesticks with tapers, surmounted by a very fair oil-painting of the birth of our Saviour, surrounded by a group of figures, a cross above the altar, and a semicircular railing enclosing the whole. There was also a large figure of the crucifixion on the left, and the walls were painted of a bluish colour. A few books were arranged near the pulpit.

From the church I. went to the churchyard, situated just above it on the rocks. There were two graves distinguished from the rest by being railed in, one, having green stakes, containing the remains of the wife of a former minister, and the other those of a governor. A broken-down head-board pointed out the spot where a fellow-countryman, a native of South Shields, and seaman on board a whaler, found his last resting-place. Several piles of stones indicated the graves of Esquimaux.

We entered a hut of the Danes, and I tried to bargain for one of the Esquimaux canoes, but they could not be induced to part with one. I exchanged two bottles of rum for three pairs of sealskin slippers, and picked up a few specimens of the purple saxifrage (*Saxifraga oppositifolia*) in flower, and a few specimens of the rocks. These are much of the same character as at the Whale Fish Islands and Disco, only that much more mica enters into their composition; indeed, in some localities giving to the soil a bright, sparkling appearance. On returning on board I again exchanged two bottles of rum for three more pairs of the fur slippers and tobacco-pouches, from a canoe alongside, and offered two sovereigns for the canoe, which was refused.

At three p.m. we got under weigh, amid sleet, snow, and fog, running the gauntlet amongst numerous bergs, stream-ice, and islands on all sides, the squadron keeping together by firing guns at intervals; and when I turned in at midnight I saw all the ships winding their way through the labyrinth of bergs. On the following day we passed Cape Shakleton, of which I took a sketch.

Thursday, 24th.—Fine Midsummer Day for these regions. Off the Devil's Thumb, Melville Bay, which, with the Sugar-Loaf, Three Islands of Baffin, and Cape Shackleton, being all in sight at the same time, I sketched from the deck—the land-floe skirting the shores, with here and there a lofty berg towering above it, and standing out in strong relief against the black surface of the perpendicular cliffs, and outside of the lead of open water, through which the squadron were winding their way, lay the "pack," a drifting mass of broken-up floes and hummocks, studded over at intervals with lofty bergs. This proved a remarkably fine day for seeing the whole line of coast to advantage. The Devil's Thumb in the early morning first appeared gradually rising in an isolated form like a lofty pillar or tower above the bank of mist.

Whilst I was standing on the port side of the forecastle viewing the grandeur of the surrounding scenery, the cheering voice of old Captain Kellett of the *Resolute* came across the water hailing us, and, with his good-natured, joking way, addressing myself by "How do you do, Sir Robert?" having, in one of his pleasant humours, when I went on board his ship to see him during his recent attack of illness, dubbed me, as he said, with knighthood.

In the evening all the ships made fast to a floe; and I landed upon the floe for the first time, and rambled over it for half a mile or so. The dingy was sent for me at midnight, the ships having suddenly to cast off their hawsers. Two of the *Resolute's* dogs were left behind on the floe to perish.

Sunday, 27th.—About two a.m. I was awakened by the report of fuses blasting the ice, and upon going on deck found that the *Resolute* had got into a "nip," and carried away her rudder; she was heeling over at a considerable angle, with hummocks of ice piled up around her sides, at the narrow point or angle of the floe, where they met in collision, causing enormous pressure. The crews of all the ships were busily engaged cutting docks in the land-floe with ice-saws in triangles, to secure them from the pressure of the surrounding floes. Our dock having given way, we had to cut a new one in the evening, which occupied some two hours' sawing through ice three feet in thickness. Our latitude to-day 75°, and longitude 60°, near the Isles of Browne, and between Allison's and Melville Bays.

Monday, 28th.—Saw a fine full-sized bear on the floe, his dirty white, cream colour contrasting strongly with the pure white of the hummocks of ice amongst which he was striding along, and which brought out his powerful form in striking relief.

Tuesday, 29th.—To-day I suggested that the main-deck ports should be opened to ventilate the ship, and, after some pressure brought to bear upon the "executive," it was very sulkily done, when the sudden bursting in of a flood of light and fresh air into my own cabin proved the greatest luxury, after above two months' subjection to a close, confined atmosphere. After dinner Toms and I went upon the floe, taking our guns, but had no sport; a solemn silence reigned around, with scarcely a sign of animal life. As we were returning we met Marshall of the *Assistance*, who had been in the year 1840 one of my boat's crew when employed in the exploration of Cumberland Bay, Kerguelen's Land, in the opposite hemisphere; like us, he was taking a ramble on the floe with his gun.

As I was getting into the dingy to go on board, the edge of the ice-floe gave way under my feet, and immersed me in a cold bath up to my waist, and I had to scramble over the boat's gunwale, with my heavy fisherman's boots on, as best I could.

Wednesday, 30th.—This morning we fell in with the wreck of the *Regalia*, whaler, of Kirkcaldy, lying alongside of a floe, where she appears to have been destroyed by a "nip." At nine a.m. Toms and I walked over the ice to the scene of the disaster, about 100 yards from us. We went on board of her. The *Assistance* was made fast close to the wreck, and her crew were busily employed wrecking. I went on board, and down to the gun-room, where I saw an egg of the Arctic gull. From the *Assistance* I went on board the *Pioneer*, her commander, Lieutenant Osborne, having sent his boat for me when he saw me passing along the floe; and from her deck I witnessed the blowing up of the wreck; it sank immediately, when casks, spars, and other gear kept rising to the surface for some time afterwards. The boats of every vessel in the squadron were soon on the spot, all scrambling for the floating remains of the ill-fated vessel. A number of Fulmar petrels (*Procellaria glacialis*) had collected together for their share of the plunder whenever oil and blubber came to the surface. Osborne showed me his cabin, where I joined him in a glass of wine and ship's biscuit; and a little before noon I returned on board, and our boat soon came back with her share of the plunder, loading our decks with firewood. A barque, the *St. Andrews*, we had seen some three or four miles distant, got under weigh in the afternoon, and stood towards us. We learnt from her that the whaling fleet were ahead of us, and in the evening we came in sight of them, thirteen sail, all ranged together along the edge of the floe, each in a dock cut in the ice, but so close together that in the distance they appeared a forest of masts rising above the icy wilderness around.

But the most novel spectacle was what followed. Their crews had all assembled on the floe, advancing along its edge to meet us, with flags flying, and marching to the tune of the old Scotch air, "Rob Roy McGregor," played by a drummer and two fifers, one dressed in regimentals, the red jacket conspicuous amongst the motley group clothed in blue and drab habiliments, and of all ages, from the boy of fourteen or fifteen to the old man of sixty and upwards. Amid the throng was a sort of triangle, and a figure dressed as harlequin in mask and cocked hat. There were about thirty colours in all, pocket-handkerchiefs included; one or two, having a star in the centre, belonging to an American whaler amongst them. They carried a boat over the ice with them, and as they passed the leading ships of our squadron gave them three cheers, which were again repeated on passing us; continuing onwards, they cheered the *St. Andrews*, whaler, after which they broke up into straggling parties, several coming on board of us to see the ship. At 10.30 p.m. we commenced cutting a dock, which was completed about one p.m. I landed upon the floe, and walked up to the dock.

Thursday, July 1st.—The whaling fleet made a move to the eastward this morning, with the exception of the *St. Andrews* and a brig lying astern of us in dock. But, not finding open water, they soon all came to again. about three miles from their former position, which was about half a mile distant from us. The *St. Andrews* was decorated by a very gay garland, suspended from her main-topgallant stay, which it is customary for whalers to hoist on the 1st of May. Our latitude to-day was 75° 9′, longitude 61° 23′ W., our present position near the Browne Isles, in Melville Bay. Nothing but level floes, about three feet in thickness, around us. But here and there, where the floes have met, forming a "nip," a line of hummocks are thrown up, and lanes of water open in places. Two small bergs alone break the monotony of the level white plain, bearing west, and about two or three miles on our quarter, and two or three more are just visible on the distant horizon, but scarcely a bird to be seen flying about. On our port side lies the *Intrepid*, and on the other side of her the *St. Andrews*, and on our starboard side the *Resolute*, *Pioneer*, and *Assistance*; and beyond them the whaling brig. The surgeon of the *Resolute* came on board this forenoon to invite me to meet Captain Kellett at dinner, as he was to dine with the officers in the gun-room to-day for the first time since his illness.

Friday, 2nd.—Overcast and gloomy, with some fine snow. In the afternoon the surgeon of the *Resolute* came alongside with the dog-sledge, to ask me to join him in a drive over the floe; our dogs, consequently, were harnessed with the *Resolute's*, but they disagreed, and would not pull together, breaking adrift from the sledge. I then harnessed the *North Star's* three largest dogs to a sledge, and drove them myself alone as far as the *Resolute*, without any difficulty whatever, in a direct course. When Lieutenant Pim landed on the floe from the *Resolute*, I gave up the whip to him, as he wished to try his hand with them; but he could not succeed in getting them away from the ship, and after meeting with a capsize, he gave it up as hopeless. The surgeon of the *Resolute* having only just before driven his sledge over the edge of that ship's dock, we sent them back to the ship, and he accompanied me on board the *North Star* to tea, five p.m.; after which we started, in company with Toms, for the whaling fleet. Before we had proceeded far, however, Toms, finding his fisherman's boots too heavy for the work, gave it up and returned to the ship, leaving the surgeon and myself to proceed alone.

Soon after passing the *Anna* brig we saw several ivory gulls on the edge of the floe, and had to cross a small opening or two in the ice before we reached the ships, a distance of about three miles from the *North Star*, which we accomplished in about an hour. There were twelve vessels in all, two of them brigs, mostly lying two in one dock, with their sterns to the floe, and their boats laying on it at a short distance.

The first ship we arrived at was the *Princess Charlotte* of Dundee, next to her the *Chieftain* of Kircaldy, then followed the *Pacific* of Aberdeen, *Rose* of Grimsby, *Jane* of Bo-ness, *Alexander* of Dundee, *Orion* of Hull (brig), *Advice* of Dundee, *Lord Gambier* of Hull, *Truelove* of Hull, *Horn* of Dundee, and lastly, the American ship, *McLellen* of New London.

On passing the *Advice*, Captain Robb, who was standing abaft, looking over the stern, spoke to us; when, after a passing remark or two about the state of the ice, he offered me a fine skull of the walrus with large tusks, as a specimen; and it proved rather a heavy one to carry a distance of three miles over the floe on my shoulders, held in front by the tusks, and I did not get on board till after midnight. We also went on board of the *Truelove*, one of the oldest of the ships, and had a long chat with her old captain, Parker, who made us take a glass of brandy and water with him. We next paid a visit to the American *McLellen*, the last ship in the tier, and had a long talk with her young commander, Captain Quayle, in his cabin. Our last visit was to the *Pacific*, where we remained with old Captain Patterson till eleven p.m., when we commenced our return to the squadron. From these old whaling captains I learnt that the Governor of Disco's name was Maltrop, that he had been a cooper, and of Upernivik, Fleishcher. Old Captain Parker told me that this should be his last voyage; he has a jolly, florid face.

Saturday, 3rd.—Received a note from Commander McClintock, inviting me to dine with him on board the *Intrepid* to-morrow, which I accepted. At 6.30 p.m., accompanied by Toms and our guns, paid another visit to the whaling fleet. Went on board of the *Alexander*, and saw the remains of the narwhal killed this afternoon from a boat with their harpoon-gun. Went next on board the *Pacific*, to see old Patterson, whose son, a young surgeon, had just had a narrow escape from drowning when falling between two pieces of ice, losing his gun. Lastly, we paid a visit to the *Horn*, the last ship but one in the tier, and her captain, Sturrock, insisted upon our going down to his cabin and joining him in a glass of brandy and water. He is a fine, jolly, florid-complexioned specimen of the seaman, like his brother-captains, Parker and Patterson, the whaling veterans, and told us that the captain of the *Alexander* was his son. He promised to look out for a bear for me. Saw a captive Arctic gull in the stern-windows or ports of the *McLellen*, picking at the glass in his endeavours to get out.

Sunday, 4th.—At three p.m. I walked across the floe to the *Intrepid*, lying on our port-quarter, and dined with McClintock, meeting Osborn, the commander of the *Pioneer*, and Mecham, the first lieutenant of the *Resolute*, who, with the officers of the schooner, made up our dinner-party. McClintock placed me on his right, and Mecham on his left hand. A good fire warmed the mess-room, and we formed a snug and cosy *réunion* in this icy wilderness. McClintock gave me a specimen of the plumbago from the mines of Sanderson's Hope, and showed me his herbarium of Polar plants, cabinet of rock-specimens, and a Ward's case of ferns.

About ten p.m., after we had our tea, a bear was reported on the floe, and I observed through a telescope from the *Intrepid's* deck that he was feeding upon something on the ice. A boat was at once manned, and shoved off in such haste that only two Minié rifles were put into it, and as McClintock and Mecham each took one, I took the helm and steered the boat for the floe, having to haul her over a small floe-piece. On reaching the floe on which the bear was, we found the distance too great for a shot, and we had to pull round a point of it, forcing the boat through a narrow channel, and so made a much nearer approach to him. Bruin, nothing daunted at being thus circumvented, very coolly and leisurely walked towards us to the very edge of the floe, looking hard at us all the while, and, sniffing the air, seemed much disposed to take the water and not await the attack, but make the charge in his own element. If such were his intentions Mecham upset them by firing his rifle, but, although only about thirty yards from us, the ball did not appear to take effect, for the bear made off immediately along the floe, but the next shot, fired by McClintock, broke the animal's spine, when, dragging his hinder extremities after him, he plunged into the water, but on finding that he could not swim, regained the floe.

By this time we had all landed, and having no gun myself, I snatched up a lance from the bottom of the boat, and on reaching Bruin with it, found him stretched out at full length on the ice, breathing heavily. As I was in the act of plunging my uplifted lance to his heart, he suddenly reared himself on his forefeet, and, with a fierce expression of eye, made an attempt to spring upon me; but at this critical moment McClintock interposed, saying that a ball would injure the skin less than a thrust from a lance, and he wished to present it to Captain Kellett; so Mecham finished the poor brute's sufferings by firing at him again, but, although within a few paces, it took three more shots from him before this was accomplished, so tenacious of life is this animal. We had next to drag him into the boat, and very nearly capsized her in getting his ponderous carcase over the gunwale. After forcing the boat through some small floe-pieces, we got on board again about eleven p.m. On measuring the bear we found him to be seven feet in length, and 539 lbs. in weight. I returned on board the *North Star* about midnight.

Tuesday, 6th.—Fine day. Walked as far as the whalers, accompanied by Toms; went first on board the *Advice*, Captain Robb, to see a narwhal, caught this morning, but the tusks, two feet in length, had been taken out, and the head destroyed. Asked Robb to dine on board with me to-day. I next went on board the *Princess Charlotte*, and invited Captain Deuchers to meet his brother-whaler at dinner. I learnt from these two old whalers the history of the voyage of the *Prince of Wales*, through Jones' Sound, either by Croker Bay or the Wellington Strait, into Barrow's Strait, the first land they made being Leopold Island, and the discovery of a cairn and footprints, as reported by her captain, Lee, and supposed, at the time, to have been traces of the Franklin Expedition, and consequently of much interest to me, as the attempt to find out this spot, and test the truth of the rumour, forms part of my own programme in the forthcoming search.

Alston, our mate, to-day made an attempt to walk over the ice to the nearest land. He started at one, and returned on board about seven p.m., but failed in the undertaking, as I had anticipated, from his having so much underrated the distance. So deceptive are distances to the inexperienced eye, amid the waste of ice and snow, as there is no middle distance to arrest the eye in these regions, an island or cape is often found to be double the distance, or even more, than the unaided eye would assign to it. When he got on board, thoroughly knocked up, he told me that at his farthest the headland he had kept in view so long appeared to him "as far off" as when he left the ship. I took a sketch of the land to-day. The two whaling captains did not come on board to dinner, owing to the very uncertain movements of the ice and ships, requiring vigilant watching. But five of them—Robb, Deuchers, Patterson, and the captains of the *Horn* and *Regalia*, paid us a visit later in the day, took wine with us, and went round the ship. After they had left us in their boat, and returned on board their respective ships, we all got under weigh together.

The Arctic squadron and the whaling fleet passing along a lane of water in the ice, the *North Star* got the start of the squadron, and stood on with the whalers, being the sixth ship in their line, the rest of the squadron getting into a "nip." We subsequently passed all the whalers, with one exception; the American ship *McLellen*, "Yankee like," took the lead. The "stars and stripes" must ever be in the van. We moved along a lane of water, just wide enough for each ship to pass in single file along it, and which eventually turned out to be only a bight between the floes. Just before we reached the end of it, the *Alexander*, which had been tracking astern of us, passed ahead of us, and we became third in the line. All the ships, with ourselves, had their crews on the floe, tracking them. The various motley groups, hauling on their lines, and marching to the time of their characteristic whaling songs and ditties, presented a most striking and animated scene. The two divisions of the ships, in all amounted to nineteen sail, ten in our van division, and nine in the rear one. On approaching the termination of this lead, the leading ships got blocked, and we had to blast the ice round us with charges of gunpowder. I went upon the floe to the bows of the American, and here we had to cut into dock.

Early in the morning, after taking a look at the fleet from the mizzen-rigging, I walked along the floe to the dock the *Princess Charlotte* was cutting, and had a chat on the state of affairs with old Captain Parker of the *Truelove*, and the captain of the late *Regalia*. It was a beautiful bright sunny night, and I remained on the floe till four a.m. The ships' companies were still at work, cutting their docks; the *Jane* of Bo-ness being in the same dock as ourselves.

The Devil's Thumb, Melville Bay. (*See page* 27.)

CHAPTER IV.

The *McLellen* nipped—Collision with our ship—Wreck of the *McLellen*—We pursue our track—Magnificent view—Cape York—In the open sea once more—We visit a native settlement—Red snow—Esquimaux curiosity—Capes Osborn and Croker's Bay—Volunteers for my boat expedition—Sight Leopold Island.

Wednesday, 7th.—This morning, about nine o'clock, the *McLellen* was severely "nipped," heeling over at a considerable angle, and huge hummocks of ice piled up and grinding along her sides. I went on the floe alongside of her; the crew were employed getting everything they could out of her upon the ice, and amongst the various things, the unhappy Arctic gull I had seen in her stern port the other day had fallen into the hands of one of the officers of the *Assistance*. The commodore made his appearance with a party of marines, four in number, and sent them on board the *North Star*, to be in readiness, if needed, to keep order on board the whaler. He had also some officers and men with him.

The unfortunate ship seemed inevitably doomed to

destruction. I went on board with her captain, Quayle, and old Captain Parker, the commodore of the whaling fleet. We went down her hold, where the damage was pointed out to me on her beam on the starboard side, which bulged inwards from the pressure of the edge of the floe without. I had a long talk with old Captain Kellett, and some of the whaling captains, who were very apprehensive for the general safety of their ships. Our own people were busily employed in blasting the ice around the stern with gunpowder. Ropes were made fast between the fore and main tops of the "nipped" ship and the hummocks to some distance on the floe, for the purpose of heeling her over, whilst the carpenters were stopping the leak. She had eight feet of water in the hold, and the ice round the ship was seven feet in thickness. Her captain, a thorough-bred sailor and fine energetic young man, acted most judiciously and coolly throughout, first of all starting all the rum to prevent all chance of his people getting intoxicated, not altogether unusual in such disasters.

Thursday, *8th.*—Weather thick and foggy, with small snow and sleet. After breakfast the *McLellen* drove her starboard quarter into our bows, carrying away our starboard cathead, and remained for the rest of the day resting in close contact with us. I stepped over our bows on board of her, and Captain Quayle told me that last night he had had great hopes of saving his ship, and had even got everything on board again. Every means had been resorted to to lessen the effect of the collision between the two ships. Our jib-boom was taken on deck, and bowsprit gear cleared away; and on board the American, her energetic captain was cutting away his mizzen-mast, axe in hand, himself—boats, davits, &c. His striking figure, and deeply-bronzed, weather-beaten face, and profusion of long, brown, wavy hair, under his broad-brimmed sombrero, made him a conspicuous figure on her dismantled deck. Just as I was leaving the *McLellen*, the commodore came on board of her, and took possession, hoisting a blue ensign and pennant, and commenced getting the stores and provisions out of the hold, and on board of us, till our decks were so lumbered we had scant room to move about them. Two more of the whaling ships, were now included in the "nip"—the *Pacific* and the brig. About noon a heavy southerly gale set in; when I went upon the floe, and from the stern-sheets of one of our boats lying there, I took a sketch of the wreck, including the *North Star*, *Jane*, and *Alexander*, and another vessel, which I finished before dinner, together with another of the boats and stores piled upon the ice from the deserted whalers. The weather very thick and dismal, and pinching cold to the fingers. After tea, I skinned the two little auk I shot. The *Alexander* is drifting on us.

Friday, 9th.—Thick weather, with snow and sleet, but a calm after the destructive gale of yesterday, which had inaugurated such a scene in the wilderness of ice around us as baffled all description, in its tragi-comic picturesqueness. Even I, to whom such wild scenes have been but too familiar in the course of an eventful life in the polar regions of both hemispheres, can never forget the deep impression made upon me by the wild, elemental war which raged around us on that memorable day. After dinner I again went on board the wreck, and found the commodore there. I met Captain Quayle as he was going over his ship's side, and invited him to come on board the *North Star* and dine with me on Sunday, which he accepted.

I had a long conversation with his first mate, who told me that he had been at the Crozets, and at my old and favourite island, Kerguelen's Land, sealing. I took half a dozen American biscuits from a cask on deck, as mementoes of the wreck. During the first watch the

R. McCormick, R.N., del. Vincent Brooks, Day & Son, Lith.

Wreck of the American Whaler "McLellen" on the bows of H.M.S. "North Star" in Melville Bay, when beset in the pack in 1852.

PAGE 40—VOL. II.

wrecked crew of the *McLellen* made a fire on the ice, where they bivouacked in front of the *Truelove* in a most disagreeable night, snowing and raining alternately; a most desolate, gipsy scene.

Sunday, 11th.—Fine day. Captain Quayle dined with me, also Lieutenant Pim and the surgeon of the *Resolute*, whom I had asked to meet him. In the course of the evening we were joined by the French lieutenant, De Bray, of the *Resolute*, Captain Walker of the *Jane*, and some others. A lovely, bright, sunny night. Captain Quayle left about nine p.m., and promised to send me a wax figure in furs of the Esquimaux. The *McLellen*, it appears, belongs to Mr. Grinnell, of New York, as chief owner. I learnt from Captain Quayle that at the Danish settlement of Julian's Hope, near Cape Farewell, the natives amounted in number to 2500, many of them having the features and hair of Europeans.

Monday, 12th.—The wreck was scuttled to-day, and on the following morning she was fast settling down.

Tuesday, 20th.—Walked across the floe to the *Truelove*, she is, I believe, the oldest of the whole whaling fleet, being ninety years of age. Parker informed me that he had met with Esquimaux at Keminsik above six feet in height, and weighing above seventeen stone.

Wednesday, 21st.—Very fine, sunny day. In the afternoon the squadron commenced tracking to the northward, the *Alexander* alone accompanying us, and we took her in tow. I gave Captain Quayle a farewell shake of the hand through my cabin-port in our hurried departure, the ice being all in motion, and a dozen or so of the whaling ships standing to the southward under a press of canvas. The commodore and Captain Kellett, with Commander McClintock and Lieutenant Sherrard Osborne, dined on board with us to-day. When I left the deck after midnight we were approaching the wreck of the *McLellen*, whose black hull just peered above the surface of the ice, of which I took a hasty sketch, with the distant view of Melville Monument as a background.

Thursday, 22*nd*.—My fifty-second birthday. Showery morning, but fine afternoon; evening foggy. After breakfast I saw the wreck astern, and the last of the whaling fleet in the distant horizon. The pack much loosened, ice honeycombed and fast decaying. Warping through narrow, winding channels between the broken-up floes. The *Alexander* still in tow astern, her crew at our capstan, warping and tramping to their airs and songs.

Monday, 26*th*.—Very fine day. Cut through the narrow neck of ice which separated us from the inshore water, following a fine, long, open lead in for the land to the left of Cape Walker, winding along between two floes amid a labyrinth of bergs. Immense flights of rotges, or little auks, were continually passing us in long streams to the southward, always in that direction, and for the most part during the night, rarely during the earlier portions of the day. Here and there a seal appeared upon the ice.

About 9.30 p.m. we were arrested by another neck of ice, three leagues from the shore, and had to make fast to a berg-piece. I walked across the neck of ice for about two miles to the northward. The scene here was one of the wildest, most sublime, and magnificent I have yet beheld in Baffin's Bay. Jet-black conical rocks, capes, and ridges emerging from a vast expanse of white, and intermingled with bergs, both large and small, skirting the coast-line, some aground and others afloat. The background exhibited an immense long line of glaciers, extending as far as the eye could reach, in one vast, mighty wreath of snow-line banked up against the horizon, Cape Walker and Melville Monument appearing on the right not far off. The ships having at last made their way through the neck of ice, about midnight we all returned on board. I remained on deck myself till two a.m. Went into the main-top, and from thence up to the crosstrees, to have a look at the lead we had just entered, when it presented a fine, open view, winding amongst both bergs and floes along the land as far as the eye could reach.

Friday, 30th.—This evening Rands, our chief quartermaster, volunteered to accompany me in my purposed boat-expedition up the Wellington Channel. At eleven p.m. the thermometer fell to 28° Fahr., the lowest we have yet had it; and we got out clear of the labyrinth of bergs.

Saturday, 31st.—Cape York bore N. 22° 40′, twelve miles distant; got into a large space of open water during the first watch, after having got through a narrow neck of ice by blasting with gunpowder, at noon. The whole aspect of the land is dreary and desolate in the extreme, the ravines and hollows between the mountains filled by deep snow or glacier, through which the black rocks project here and there, and have a very striking effect as they peer above the white, snow-clad surface. One snowy, cone-shaped peak, inland mountain I sketched.

Sunday, August 1st.—Fine day. This morning we found ourselves in an open sea, with scarcely a bit of ice to be seen, having cleared the margin of the pack off Cape York, of which, with Bushman Island, I took a sketch. A boat was sent on board the *Alexander* to invite Captain Sturrock to dine with us, at my own suggestion, before we parted company. Soon afterwards we cast off the hawser, and both ships made all sail into the bay on the south side of the cape, where a group of some seven or eight Esquimaux, with half a score of dogs and a small sledge, had come down to the floe-edge to meet us.

The letter-bag, having been closed, was sent on board the *Resolute*, and in the boat I took a passage for the shore, about noon. The *Intrepid's* bow was resting on the edge of the land floe. Captain Kellett, on the floe, bargained with the natives for a small sledge, for which, they received from him a couple of planks, a saw, and some knives and scissors. The oldest of the natives, the "Arctic Highlanders" of Ross, was goodnatured-looking, and wore goggles, the frames being made of wood; he was accompanied by another man, and two or three boys, all dressed in sealskin jackets and bearskin trousers and boots. Although dirty in person, they were all very ruddy and healthy-looking. We all walked on to the settlement; amongst the party were Captains Kellett and McClintock, with the French lieutenant, De Bray, and Nares, mate of the *Resolute*. Our course over the floe was intersected by numerous cracks and pools, frozen over. We leaped over some, and slided over others, quickly, which were barely thick enough to bear our weight for a moment.

After proceeding for about two miles we reached the encampment, consisting of three small sealskin tents, on the side of a hill above the floe. All surroundings, as usual, of a filthy description; a small sealskin laid out to dry, and the entrails of another with numerous skins of the little auk, scattered about. A bitch, with some half-dozen pups, lying outside the tents. The rotges were very clamorous on the hill-side above, and I at once started off across a spongy, moss-covered bog, and up a *débris* of fragments of plutonic rocks, in search of their eggs and young, but found neither. But before I had got half-way up, the recall was made, and after picking up at the highest part of my ascent some of the much-celebrated red snow of Sir John Ross, a few specimens of the rock-formation and plants, I returned through the settlement. The women and children were much amused and attracted by a red night-cap I had in my hand, containing my specimens; they ran after me, laughing and talking most heartily and joyously, feeling my

R. McCormick, R.N., del. Vincent Brooks, Day & Son, Lith.

Cape York, North Greenland.
The Esquimaux Settlement of the Arctic Highlanders.

PAGE 44—VOL. II.

pockets, till after a hurried shake of hands, I took my leave of these goodnatured and unsophisticated children of nature.

I overtook Lieutenant Pim, and we walked together till we reached the floe-edge, where, since I left her, the *North Star* had made fast. On our way we discussed our own somewhat anomalous and difficult positions, held in this expedition. About 2.30 p.m. I returned on board, where I found the two young Esquimaux boys, fine, healthy, lively, intelligent lads, delighted with everything they saw, especially amused with an eye-glass I wore round my neck, looking through it with excited curiosity, at its magnifying power, as did also their brethren on shore. I exchanged with the eldest boy one of my scalpels he took a fancy to, for one of his rude knives, made from an old iron hoop. Toms and I showed them our cabins. They, very anxious to obtain wood, pulled out one of the cask-staves, &c. The captain of the *Alexander* dined with us. In the evening I took sketches of Cape York, Dudley Diggs, and the Crimson Cliffs, only a pale pink ravine and patch or two of it, now appearing. The white pack margining the horizon all round.

The modern Esquimaux are supposed to be descended from the artistic reindeer hunters, Palæolithic or ancient European cave-men, who were the representatives of the Neolithic Age, during the Pleistocene Period, when the river-drift men, the oldest race, probably, first appeared, towards the close of the Quaternary or Glacial epochs.

The vast glaciers bounding Melville Bay, which are said to be moving at the rate of one hundred feet annually, separate this tribe of the pure race of Esquimaux from the half-breeds of the Danish settlements, and having no canoes, all communication is cut off between them. Their only mode of travelling is by sledge, drawn by a fine breed of dogs, with soft, silky hair and very symmetrical forms, though small in size.

Monday, 2nd.—This evening we were standing over in the direction of Jones Sound, with a strong breeze, amongst heavy pieces of ice in streams, a thick fog obscuring the land; lost sight of the *Alexander.* During our passage up Lancaster Sound we had a dead beat, and did not even get a sight of the southern shore, on our way to Beechey Island. We sighted Cape Horsburg, Hope's Monument, Capes Osborn and Warrender, with Croker's Bay, and had a fine view of the top of this bay, well up it, and most distinctly saw the land "continuous" all round, in every direction; no passage certainly exists from it. The remarkable trap-dykes, about four in number, intersecting a truncated, conical mountain, with a fine specimen of a glacier flowing between the rocks to the sea, as shown in the opposite page, appears near the entrance to Lancaster Sound.

Saturday, 7th.—Rands, our quartermaster, whom I have made coxswain of my boat, told me last evening that Millikin, one of our most active young able seamen, had volunteered for our boat, and that this morning, Burns, another able seaman, had offered, and before the close of the day three more came forward, completing the number for my boat's crew. On the following day, Sunday, 8th, I saw the land, on the south side of the sound, for the first time, Leopold Island, on port bow, table-topped, and terminating in a bluff headland, striated here and there with snow. Passed Hobhouse Creek, a league and a half from the coast, which is of limestone formation; the face of the steep cliffs having a remarkable and very symmetrical buttressed appearance, tier above tier. I took a sketch of Maxwell Bay, embracing Capes Fellfoot, Herschell, and Hurds. Whilst walking the quarter-deck with Alston, our mate, during the first watch, he renewed his offer, made once before, to serve under my command in my boat-expedition up the Wellington Channel.

Remarkable trap dykes and glacier, Lancaster Sound. (*See page* 46.)

CHAPTER V.

Beechey Island—Off Caswall Tower—Cape Riley—A bear chase—A cairn and graves—Wellington Channel—We inspect the *Mary*, left by Sir John Ross—I apply for leave to start on my expedition whilst the channel is open—The commodore interposes obstacles—My plans thwarted—Our ships disperse—My boat ready for starting.

Monday, 9*th*.—A very fine day; this morning found us off Caswall Tower, in Gascoigne Inlet; I took a sketch of it. About one o'clock a boat was lowered, in which I took a passage to land on Cape Riley, accompanied by my young friend Toms, for whom I asked a passage. On landing we found a memento in a cylinder, beneath a cairn and a staff. We re-embarked and went upon the floe which still fills up Erebus and Terror Bay, bounded by Cape Riley to the eastward, and Beechey Island westward. From this the boat again took us to the foot of Cape Riley, and on landing, we at once commenced the ascent of Cape Riley, up the south steep face of the cliff, over a loose *débris* of limestone rock, abounding in fossils, more especially "producta," of which I picked up several fine groups in fragments of the limestone, and a few specimens of the purple saxifrage, and arctic poppy, in bloom. We reached the summit in twenty minutes, walked over its flat table top to the opposite end, overlooking the bay beneath; and beyond that Wellington Channel quite clear of ice; what an encouraging prospect this held out to me, could I but then have started in my boat without further delay; but unhappily for me this was not to be. We descended the mountain nearly at the part from which we made the ascent. Walked across the floe to the ship, already made fast to the edge of it, and got on board at four p.m. to dinner and tea. In the evening I crossed over the bay-floe, to the long, low, shingle ridge, dividing Erebus and Terror Bay from Union Bay, accompanied by all our Esquimaux dogs, whom I could not shake off.

At seven p.m., when about halfway across, a large bear hove in sight, coming from the upper end, and galloping along at a rapid rate towards me. I at once prepared for action by loading both barrels of my old double gun with ball, but on the dogs getting sight of him, they started off, in spite of me, in pursuit; and consequently I lost all chance of a shot at him, as he immediately made off for the hills on the opposite side of the bay, followed by the dogs in full cry, till I lost sight of them all, over the ridge. Another bear appeared near the three graves of Franklin's crew, but he also disappeared over the ridge to Union Bay. On reaching this ridge, I found a bamboo pole above a cairn of fifty empty preserved meat canisters, having a tin cylinder attached by a piece of weathered white line. I next visited the three graves: the headstones were painted in white letters on a black ground, all faced the east, the *Terror's* being nearest the bay, and only covered by shingle; the *Erebus's* two had over this a few large slabs of limestone.

Three boulders of granite were lying near, and one very large mass a little higher up on the ridge. The spot wore a melancholy, sombre air of utter desolation and loneliness. Having by this time gathered my dogs around me again, I ascended the steep face of the craggy cliff above the graves, by which I was rewarded with a fine view of Wellington Channel, quite open and free from ice beyond Cape Spencer. After walking for some distance along the summit of Beechey Island, I descended on the north side, and here I again parted from the dogs, who did not return to the ship until some twenty minutes after I did, with one exception, and much to his credit, my dog Erebus never once left my side throughout the whole excursion; my other dog, Terror, was missing for some hours. On my return, I found the ship's company employed in cutting a dock, and I got on board by means of the bowsprit ladder-rope. At about 2.30 p.m. the *Resolute* and *Intrepid* joined us, and I ultimately turned in at three a.m.

Tuesday 10*th.*—Fine. Lieutenant Pim came on board to ask me to join him in a ramble to Union Bay, to see Sir John Ross's decked boat, the *Mary*, of about twelve tons, left by him under Cape Spencer. We took our guns, and after passing by the graves and cairn, descended the embankment of snow to the other side of the shingle ridge, into Union Bay. Here we saw the enormous footprints of the bear I fell in with last night, in the vicinity of the graves. About a mile from the ridge, and four miles from the ship, we found the *Mary*, and here we met old Captain Kellett, and two of his officers, Lieutenant De Bray, and the master. The mast of the yacht was still standing, and we found a few barrels of provisions with a memento from old Sir John Ross, and my former old boating companion, Lieutenant Phillips.

I had a long conversation with Captain Kellett about

the expedition, and my own boating voyage, stating to him how anxious I was to get away from the ship, up the Wellington Channel, whilst there was yet open water. But he told me that the commodore had given no orders to him whatever on the subject. We walked the whole of the way across the floe, back to the ship together, and he was very communicative and chatty. I learnt from him that my little dog Terror was found lying near the graves by one of his ship's company, and was brought on board between six and seven o'clock this morning.

Wednesday, 11*th*.—I saw a great number of white whales pass this morning. On the arrival of the *Assistance* this evening, I at once wrote a letter to the commodore as follows:—

"H.M.S. *North Star*,
"Beechey Island,
"*August* 11*th*, 1852.

"SIR,—I have the honour to inform you that I have a boat's crew of volunteers, ready to accompany me in a boat and travelling expedition in search of Her Majesty's ships *Erebus* and *Terror*, under the command of Captain Sir John Franklin.

"I am, sir,
"Your most obedient, humble servant,
"R. M'CORMICK,
"Surgeon, R.N.

"To Captain Sir Edward Belcher, C.B.,
"H.M.S. *Assistance*,
"Commanding Arctic Expedition."

Thursday, 12*th*.—I had told my marine servant, Clark, that it was in contemplation to effect an exchange between some of our own crew and those of the other ships, for their invalids, and that I had reason to believe he would be one of the number. I asked him how it was that he had not been a volunteer for my boat; his reply was that the first he had heard of it was last night, and moreover that he should like to have accompanied me, adding, that I might have had the whole ship's company, had I needed as many volunteers, to choose from. I consequently told him that I would apply for him. Duncan, another of our able seamen, volunteered for me the other day.

After breakfast I went on board the *Assistance*, and met the commodore on his quarter-deck, with Captain Kellett, and asked him at once what boat I was to have, when he replied, "One of the light ones on board of us." This I was obliged to object to, as quite unfit for the nature of the service I was going upon, for the first piece of ice we came in contact with would have crushed her like a nutshell. He then said I must have a suitable boat, and that on his return on board he would arrange about it, and send for me. On my return on board my own ship, the ice under my feet giving way alongside, I was submerged to the waist, and had to change all my clothes, and found that my watch was injured by it. This will prove a great loss to me for the rest of the voyage. My servant, Clark, told me I had another volunteer, an able seaman, Steel. This evening I arranged my cabin preparatory to my departure in my boat, and did not turn in till midnight.

Friday, 13th.—At eight a.m. the commodore sent for me, and very insidiously attempted to shake me from my purpose in going away this autumn in my boat; when, on finding that I was firm in my resolve, by way of trying my nerves, he threatened me with calling a council of the captains of the squadron to have their opinions first. So in about an hour afterwards he sent for me again, when I found him with Captain Kellett and the commander of the *North Star*, on the quarter-deck of the *Assistance*. Taking us all down to his cabin, he commenced by asking me in an authoritative tone of voice what my plan was, at the same time intimating that it was his intention to reserve to himself entirely the exploration of the Wellington Channel. This I fully expected from him; but I was not prepared for what followed, even from him, when he endeavoured to make it appear to the others that I had, in a former conversation with himself, on the passage out, expressed a wish to confine myself to the search of the sounds at the upper part of Baffin's Bay. The truth was, I had had no previous conversation whatever with himself on the subject. Of course, in the presence of my superior officer, I could only reply to this, by a calm negative, and that my views had been directed more especially to the Wellington Channel, and that the Hydrographer, Sir Francis Beaufort, had himself made arrangements for my first searching that quarter and Baring Bay, with the view of setting at rest the much-mooted question, as to the position of a cairn and footprints, said to have been seen by Captain Lee, of the *Prince of Wales* whaler, in the upper portion of Jones Sound. His answer to this was that Captain Lee himself had admitted before we left England that there was no such thing. The discussion, which naturally was by no means of a conciliatory character on either side, ended by the commodore giving me a letter, with a request that I would give him a written reply—emphatically adding, as I left his cabin, "You cannot go away till the *North Star* is cut into winter quarters." The others present said little, and I returned on board my ship, disgusted more and more with the proceedings of this ill-omened, ill-starred expedition.

Saturday, 14th.—Fine day. Last evening there was a grand champagne supper on board of the *Assistance*, on parting with the rest of the squadron. Her surgeon came on board to ask me to join in the revels, which however, I declined; and having turned in at eleven p.m., I was awoke about midnight by the French officer, De Bray, and Groves, one of the mates, coming into my cabin with the message from good old Captain Kellett, begging me to come over—well meant on his part—but the temptation under existing circumstances was not strong enough.

About eight o'clock this evening all hands from the ships were piped up to assemble on the floe, when the commodore in the centre of the circle formed gave a parting address. But as an oration the failure was complete. He read a prayer afterwards from a printed paper, and said, he should be glad to shake hands with any one present. His entire breach of faith, and generally uncourteous line of conduct, pursued towards myself, quite forbade my coming forward and offering my hand. I accompanied Pim on board the *Resolute*, who, like myself, did not go near the commodore. We sat at the gun-room table, chatting with the officers of the *Resolute* till one bell after midnight. De Bray, who sat next to me at table, with the clever tact of a Frenchman, took my watch to pieces, and put it together again, on the plate before him, to ascertain what injury it had received the other day, when immersed in the salt water with me, as it had stopped ever since; and after some two hours' examination, found a small nut on the plate in the works broken off by the corrosion from the salt water, rendering it altogether beyond repair out here.

Sunday, 15*th*.—Fine day. I made a final effort to get away in my boat by writing a letter "on service" to Captain Kellett, now senior officer, to induce him if possible to give me permission to leave immediately, but he did not appear to feel himself justified in departing from the commodore's orders only so recently given. This he gave me in writing, in order to exonerate me from all blame in the delay with my friends at home.

On going on board the *Resolute*, he received me with his customary kindness, and I had a long chat with his officers on deck, by whom I was surrounded, wishing me every success in my enterprise. At two p.m. the *Resolute*, with her tender the *Intrepid* (McClintock), sailed for Melville Island, the *Assistance* and *Pioneer* (Osborn) having proceeded up the Wellington Channel last night.

Tuesday, 17*th*. I have been all day employed in the equipment of my boat, and making a demand for stores, provisions, &c., ordering Rands to muster my boat's crew on the quarter-deck, where I explained to them that I had selected them according to their seniority in offering themselves as the fairest method, all being equally good men, and well adapted for the service. My own servant, the marine, was the only exception made. Had a sealskin suit, with fur gauntlets, served out to-day, and took up two purser's shirts.

Wednesday, 18*th*.—Blowing a gale of wind with a fall of snow during the night, mantling all the hills with white, whilst icicles were suspended from the rigging and spike-plank. Weather very squally all day. The party sent to bring round the *Mary* from Union Bay did not return with her till four o'clock this morning. I was employed all day in completing the equipment of my boat. Nothing but the wind being dead ahead and blowing a gale prevented me from starting this evening, but as nothing could possibly have been gained by it, I have deferred my departure till to-morrow morning after breakfast. The *Mary* yacht is now lying astern of us.

Thursday, August 19*th*.—I left the *North Star* at eleven a.m., accompanied by Toms and another of our messmates, over the floe to my boat, lying alongside of the floe-edge, about a quarter of a mile distant. They both gave three hearty cheers as I shoved off, which was responded to by our people working at the triangles with ice-saws in cutting a passage for the ship through the ice. We returned them three more as hearty from the boat, and, hoisting the lug-sail, ran before a fresh breeze alongside Lady Franklin's brigantine, the *Prince Albert*, which most opportunely hove in sight just in time for me to communicate with her before I departed on an expedition upon which I had so long set my heart, and gone through so much anxiety, and hopes long deferred, to attain.

Sir John Ross's yacht, the *Mary*, Beechey Island. (*See page* 54.)

CHAPTER VI.

I return from the boat expedition—On board the *North Star*—Ship driven ashore—Severe weather—Blasting the ice—Cape Spencer—The last of the sun—Set a fox-trap—An Arctic raven—Christmas Day—The old roast beef of England—Letting in 1853—Hard weather and good cheer—First sunshine for three months.

Wednesday, September 8th.—I returned on board my ship, at 8.30 p.m., after an absence of three weeks, a period of stormy, boisterous weather, such as I have rarely encountered. It was one continued succession of heavy gales, with scarcely a sight of the sun. On reaching Union Bay I landed under Cape Spencer, to give my boat's crew a little refreshment after the toils of the day, when a boat under sail—the second gig, in charge of the second master and five hands—came from the *North Star* bound up Wellington Channel, with a *cache* of provisions to be deposited at Cape Osborne for the use of the spring sledging parties. I cautioned him not to attempt to proceed further than Griffin's Bay, or he would run a great risk of losing his boat, as I had myself remained till the very latest moment the advanced season and formation of young ice would permit of.

R. McCormick, R.N., del. Vincent Brooks, Day & Son, Lith.

H.M.S. "North Star" on shore in "Erebus" and "Terror" Bay, in the year 1852.

PAGE 56 – VOL. II.

Indeed, he returned at six p.m. on the 13th, unable even to reach Griffin Bay, and had to deposit his *cache* on this side of Cape Bowden, from the rough weather and ice he met with. On returning on board, which I did through the narrow strait between Erebus and Terror and Union Bays, which had opened since my departure, I found my shipmates all on deck to receive us, the gun-room steward having sounded his bugle as we neared the ship, and which I answered by firing off my gun. After going down to the gun-room, and having had a shave, wash, and a cup of tea, I retired to my cabin to read two letters from home, and one from my good friend John Barrow of the Admiralty, enclosing me the last "Navy List," and which were put into my hands by Toms. The *Isabel*, Captain Inglefield, had only left at three o'clock this morning for England, having on board my old shipmate and friend at both the Poles, Abernethy.

Tuesday, 28th.—Ushered in an eventful night for us, a large quantity of heavy ice off the entrance to the bay, drifted in with an equinoctial gale from the southward. During the first watch some heavy floe-pieces struck the ship with a loud, grating noise, as they swept along her sides. The cable surging heavily, the anchor dragged, and she drove on shore, but fortunately took the ground easily, on a soft bottom at eleven p.m. The evening had been beautifully fine, with a full moon, and consequently spring-tides and nearly high water. The ship heeled over at a considerable angle, rendering it very difficult to walk the deck. The whole bay became filled with ice, and one extensive floe-piece had driven us on shore in from twelve to fourteen fathoms.

This was one of those untoward events I had myself long anticipated, and had repeatedly called attention to, in consequence of the position the ship had been so injudiciously placed in, by anchoring her in the very line of drift which the ice would follow in either its ingress or egress from the bay, and added to this, a strong current sets through the narrow strait dividing it from Union Bay, disturbing the formation of the young ice, and thus keeping the bay open late in the season.

The only secure anchorage, and there the ship would have been as safe as in a dock, was much higher up, near the top of the bay, under the shelter of the high land above the graves, in the vicinity of what was doubtless the Franklin encampment, and the position of the *Erebus* and *Terror* when they wintered here. This would not have escaped the observation of such an experienced seaman and navigator as the immortal Franklin, being quite out of the line of drift.

Finding that the old ship settled herself quietly down on the soft bottom, and the worst having happened, for, as I before said, the event was one far from being unexpected by myself, I turned into my berth at about two a.m.

Wednesday, 29th.—Blowing a hard, southerly gale all day, and at intervals in very heavy gusts, accompanied by small, drifting snow, and a thick mist obscuring all surrounding objects, and the bay filled with ice. The ship last night had righted during the rise of the tide, to fall again to her bearings at low water, having only six feet of water alongside. We could not move about the decks without the aid of "life lines" and battens across. The poor dogs came off from Beechey Island to-day, half starved, and perishing with the cold.

Saturday, October 2nd.—Upon going on deck this morning, a dismal prospect met the eye on all sides, snow falling fast in large flakes, a thick mist surrounding all, and shrouding every object in obscurity. The deep snow which had fallen during the night covered the decks and rigging. The ship down to her bearings on the bank of shingle, at an angle of about 25°, beset all round with ice and snow, and winter fast closing in, constituted anything but a cheering prospective. The cabins on the port-side are now scarcely habitable, with the thermometer at zero; in my own, the vapour freezing in crystallized masses on the heads of the metallic bolts, and the toes and fingers felt as benumbed and pinched as if I had been sitting on the upper deck, the "Sylvestre stove" ceasing altogether to act on my side of the ship, and transmitting all its heat along the opposite or starboard side, being so much elevated above the other; so that, in short, we have an Arctic climate on the one side, and a sub-tropical one on the other. My only hope is that as the ice continues to form and becomes thicker under her bottom, it will gradually raise her up ultimately, upon an even keel, which will have the effect of bringing about an equilibrium of heat from the Sylvestre, and a horizontal position to the decks to move upon.

Sunday, 10th.—Overcast and cloudy, but fine calm day induced me, accompanied by Toms, to walk over the floe to the cairn and the graves, from which we ascended to the summit of Beechey Island, by the snowy ravine just above the graves, and continued our walk across the whole length of the island to the frowning precipice overhanging the entrance to the berg, proceeding from west to east, having Cape Riley opposite to us. We descended the steep face of the cliff by sliding down a bank of snow just opposite the ship, and got on board at our dinner hour, two p.m.

Friday, 15th.—Ever since the ship has been ashore, I have continued to sleep in my cabin, but almost frozen into an icicle at times, the atmosphere of it as chilly as an underground vault, and more resembling a frozen cave than anything else. Poor Alston, my next neighbour, decamped from his den to the starboard side for the benefit of the Sylvestre, but Toms, whose cabin was next to his, like myself held on to the last.

Monday, 18th.—Crew employed as usual on the ice

outside the ship, cutting and blasting the floe under the starboard quarter, now a new freak, amongst the varied vacillating and impotent manœuvres that have been in progress for some time past for getting the old craft afloat again; and but for the strength of her doubling, one of the heavy charges of gunpowder exploded right under her counter, might well have blown the timbers of her stern-frame out of her. It all but capsized me, chair and all, as I was seated in the gun-room, and set all the crockeryware dancing a hornpipe on the sideboard; her entire hull was shaken from stem to stern. The whole might have been viewed as a mere ludicrous farce, but for the inconsiderate exposure of the unlucky crew—almost daily when possible—in this inclement season of the year to intense cold and fatigue, till darkness put a stop to their disheartening labours, for the ship's position has not been benefitted an inch, a position such as ship never before wintered in, in these regions. The melancholy moaning of the poor famishing Esquimaux dogs was in keeping with the whole, and formed a climax to it.

Saturday, 23rd.—The finest day we have had this month, and I took advantage of it to make an excursion to the summit of Cape Spencer. Starting at ten a.m., I crossed over the floe to Union Bay, ascended the highest part of the cape; walked over its flat top, towards Wellington Strait—affording me a fine prospect of Cornwallis Land and North Somerset, with the channel between them frozen over. Large boulders of reddish-coloured granite were strewed over the plain surface of the summit. On my return, I visited the cairn at the bluff-end of the ridge, overlooking Erebus and Terror Bay. It is about seven feet in height, and on the very edge of the cliff. Here I descended by a remarkable block of limestone, projecting abruptly from the face of the cliff. This excursion occupied me five hours.

Sunday, November 7th.—When on the floe, near Northumberland House, I had a last sight of the sun for this season, only half of its disk now appearing above the horizon, which was clear of clouds, and a fresh breeze blowing at the time. The thermometer 20° below zero. I had both my cheeks and chin, with the right ear, frost-bitten. The skin felt rough, elevated, corrugated, and hard, giving a horny sensation to the touch. I at once rubbed the parts with snow, and on my return on board the face became swollen, with a circumscribed redness on each cheek, and sensation attendant on a burn. In the evening the surface rose in a blister, with serum beneath, followed by desquamation of the cuticle, but unattended with much reaction or pain. I attributed this sudden attack to my having incautiously, and with too much confidence in former habits of exposure, gone at once out of the gun-room, from a temperature of 50° above zero to 20° below it, without any additional covering either to neck or face, having only my regulation cap on the crown of my head, and this proved too much for the skin to resist; and I thus, on the very day that the sun took leave of us, got frost-bitten in remembrance of it.

Friday 12th.—During the first watch I saw the aurora for the first time this winter, in the form of two pale yellow rays, passing a little to the westward of Ursa Major. Our people employed in banking up snow round the ship's sides, in cold and wet, out of fear and apprehension that the ship might fall over on the starboard side at the next high tide; because the ice around cracks and grinds somewhat loudly during the expansion in the process of congelation.

Tuesday, 16th.—As I had all along foreseen from the very first, the ship, from the gradual accumulation of ice beneath her keel, has now attained her normal upright position. Her sides, housing, spars, and rigging all white with the covering of ice and snow. The "Sylvestre" stove now acting equally on either side.

Sunday, 21st.—To-day I found the tip of my right ear painful on exposure to the wind on deck, although a fortnight has elapsed since I was frost-bitten. The new skin had formed on the face last Sunday, but the redness and scar, from its repeatedly peeling off, had not disappeared until to-day.

Tuesday, 23rd.—To-day I set my first fox-trap on the floe beneath Cape Spencer, taking it there on a sledge drawn by four of the ship's dogs. I baited it with the remains of the young gull which died on board, and a piece of preserved meat. On the following day, Wednesday, 24th, when visiting my trap, I heard the hoarse croak of raven distinctly several times, but it was yet too dark for me to see the bird, whose melancholy cry broke so strangely on the death-like stillness and solitude of these wastes of snow and scenes of desolation around me. On reaching the base of the precipitous crags of Beechey Island, however, the sable and sombre-looking bird itself appeared, its black plumage shining with a metallic lustre, in strong relief against the pure white of the snow, as it rose from the rocks. I had my gun with me, but "Ralpho" was far too wary to permit me to get within shot of him.

I resolved now to complete the circuit of the island, which I had not hitherto accomplished. The cliffs around me for the most part steep and precipitous, the snow forming angular-shaped masses at their bases; the floe outside very hummocky, and covered with deep snow. When about halfway round, I found the ice forced up under great pressure, in some places to the height of from thirty to forty feet, and the snow in the hollows between, waist-deep, and so soft as to render the progression over it extremely fatiguing. I had got through about two-thirds of the circumference before

Cape Riley opened, and passing round the perpendicular, steep bluff-head forming the west side of Erebus and Terror Bay, I returned on board at two p.m., having started at nine a.m., and must have gone over, at the least, some half-score miles of ice and snow. The thermometer ranging at about zero.

This excursion round Beechey Island subsequently became a favourite one with me during the winter months, and when walking round without any stoppages or delays, I usually accomplished it in about two hours, so that the entire distance round may be fairly estimated at seven or eight miles.

Thursday, December 2nd.—Fine day; thermometer 31° below zero. I left the ship at noon, accompanied by my dog Erebus, on a visit to my fox-trap, which I had removed from Cape Spencer to the west end of Beechey Island. On approaching the trap, I noticed that the door was down, and on reaching it heard the low, gruff bark of poor white reynard, who had been caught since my visit to the trap yesterday. On opening the lid, he made an attempt to bite me, but my stout fur gauntlet saved my hand from his sharp teeth. Hauling him out, I soon put an end to his hopes and fears by pressure in the region of the heart. He was a male, and by far the finest fellow as yet caught, weighing eight pounds. I skinned him the same evening, and gave Erebus the carcase next day, on revisiting my trap, a walk of two miles.

Tuesday, 7th.—Blowing fresh, with thick weather; revisited my trap; thermometer at zero. Accompanied by Erebus; found another fox, with marks of age, in the trap.

Saturday, 25th.—A fine Christmas Day. I took my now customary walk round Beechey Island before dinner with Erebus, going out by Union Bay, and met Toms near my fox-trap, who had come in the opposite direction. We returned together just in time for dinner: roast beef, kept since we left England, a boiled leg of mutton, and a goose, the latter none the better for keeping so long, followed by the orthodox plum-pudding, and a glass of ale.

Before I retired to rest, I went over the ship's side upon the floe, and found a bright full moon shedding its soft light over the floe, from a clear blue sky; night calm and milder. Within the Arctic Circle the moon neither rises or sets; for ten days together shining continually as it circles round the horizon—in some measure making up for the absence of the greater luminary, the sun, during the long, dark winter's night.

Sunday, 26*th.*—Another fine, calm, clear day. There was an eclipse of the moon before I was up this morning, consequently I missed seeing it. Accompanied by Toms, I visited my trap round Beechey Point, and we had but just passed the *Mary* yacht, returning, when I noticed my dog Erebus scrambling down the angular slope of snow reclining on the base of the cliff, with something in his mouth, and on my making a signal to him, he dropped it at his feet, standing over it till I got up to him, when I found a fine ptarmigan, in its pure white winter plumage, killed, but otherwise not in the least injured for a specimen. Any of the other dogs would have devoured it instanter, in spite of my presence. It was a solitary bird, no trace of any others. I still have it in my collection, with my poor old dog himself, but, alas! it is only his inanimate stuffed form.

Friday, 31*st.*—The last day of the old year is now closing in upon us, and I took my favourite walk round Beechey Island before dinner, accompanied by my faithful Esquimaux companion, Erebus, proceeding by the Union Bay ridge. Baited my fox-trap afresh, as I passed it in mid-distance. About halfway round the island, met Toms at his trap, and we returned together, passing by the *Mary* yacht. He found in the stomach of his fox he skinned to-day some feathers of the very gull I had baited my own trap with, a striking evidence of his being the identical animal caught in my trap, and had eaten his way out, as I had discovered by the damage done to my trap; he had afterwards entered that of Toms. At midnight the bugle and violin were being played at the gun-room door. I went on the floe afterwards to have a look at the night, which was misty and overcast, before I turned in at two a.m., and now commences a New Year in our monotonous life.

Saturday, January 1st, 1853.—The New Year has opened upon us with the severest weather we have hitherto experienced, blowing a hard, southerly gale, misty, and overcast, with snow-drift. I gave my two dogs, Erebus and Terror, the Arctic fox for their New Year's dinner.

To cheer their spirits under the depressing influences existing in our little community, I gave to Rands, the late coxswain of my boat, a bottle of port wine, as he does not drink spirits, and a bottle of brandy for him to distribute amongst our boat's crew, and another of rum for four others who had been early volunteers for my expedition; a second bottle of rum to Harvey, our old boatswain's mate, to be shared by three other old Polar hands—the armourer and sailmaker, with Page, the old quartermaster, and the dog-keeper, who looks after my two dogs. All these were trustworthy, steady, temperate men I could rely upon to make a proper use of such a welcome addition to their enjoyment of the festivities of a season never to be forgotten by us mortals. These men, I knew well, would share everything with their messes. I closed the day by looking over the Arctic blue-books, as an evening's amusement, till I turned in at midnight.

Tuesday, 4th.—Temperature fallen to 42° below zero.

At eleven a.m. I visited my fox-trap, and here I made the attempt to scale the steep cliff immediately above it, but after much laborious exertion, on attaining the first gap, and surmounting a steep, overhanging crag, I unexpectedly found another steeper ridge of snow above this. This, too, after some risk of being hurled to the bottom headlong by the slightest false step, I, by a great effort gained the crest of this crag also; but here, being compelled to take off my gauntlets, the better to grasp the projecting point of rock to hold on by in such a precarious foothold, my naked hands lost all sensation, my gun slipped from my grasp down the declivity, and on casting a glance upwards, the acclivity still above me appeared steeper even than the one I had just surmounted, and as from my position I could not see the summit, I followed the course of my gun, and on picking it up slipped downwards for some paces, but fortunately my fall was arrested by a projecting crag of the gap, which I laid hold of, and from thence beat steps in the hard, frozen snow with the butt end of my gun to the bottom. On replacing my gauntlets, I found all the finger-ends of both hands smartly frost-bitten, in the brief space of the minute or two they were bare; skin white and hard, with a burning sensation, followed in the evening by large blisters filled with serum.

My faithful dog Erebus, who followed, or rather preceded, me to the highest portion I attained in the ascent, though very reluctantly, apparently, with more foresight than his master as to the practicability of the attempt, made a rapid descent before me, whining and exhibiting marked symptoms of impatience and dissatisfaction—in short, unmistakably rebuking me. Found Toms at his trap, and we returned together over the shingle ridge and bay floe on board.

Tuesday, 11th.—I walked round Beechey Island, now become almost a daily routine with me. The thermometer fallen to 51° below zero, but being a fine, calm day, without feeling any unpleasant effects from it, notwithstanding, I was three hours absent from the ship, exposed to this very low temperature. I shifted my fox-trap to the low slope where Beechey Island is ascended from the south. My dog Erebus seemed to suffer in his feet, having frequently to roll himself in the snow-drift to restore animation; and I felt it pinching to the hands if I took off my gloves only for a few seconds. The continuous loud cracking of the floes to-day resembled the sound produced by striking a hollow, metallic globe.

Thursday, 20*th.*—The anniversary of the fearful gale we encountered in the *Erebus* in the vast pack in the Antarctic seas in the year 1842. I left the ship to pay a visit to my fox-trap at ten a.m., accompanied by my two dogs. It was a boisterous day, blowing in heavy squalls, with a blinding snow-drift. I had put on my sealskin dress, jacket, trowsers, cap, helmet, and gloves, or gauntlets, for the *first* time during the voyage, and my dogs, noticing the change, capered round me, delighted beyond anything with the metamorphosis I had undergone, thinking, I suppose, that I had turned into an Esquimaux.

After clearing the snow out of my trap, I walked from the end of the shingle spit round the top of the bay, accompanied only by Erebus, Terror having left me, a by no means unusual occurrence with her. When I reached the second ravine, opposite to the ship, so thick was the mist and fine snow-drift, that I could neither see the ship nor Beechey Island, the moon appearing only faintly through the fog over the ridge. My sagacious dog, Erebus, showed much impatience with the drift in his eyes as he headed to windward, every now and then turning round at intervals to watch and direct my steps towards the ship, of which, though invisible to both of us, he evidently recognized the position better than I did.

Wednesday, February 2nd.—As I had anticipated that possibly, according to the laws of refraction, the sun might make its appearance above the horizon to-day, I started from the ship at eleven a.m., accompanied by my almost constant companion, Erebus, and a messmate, for the summit of Beechey Island, ascending by the snowy pass above the graves. On reaching the south side of the island, the glorious orb of day appeared for the first time for the last three months, refraction having brought its disk above the hills of North Somerset, on the opposite side of Barrow's Strait.

I first observed what appeared like the gilded margin of a cloud, from which the upper limb of the sun soon emerged. This was about the hour of noon; the surrounding horizon was suffused with a reddish light, and there was also the same pink-coloured aurora flush in the light clouds to the north as seen yesterday.

We descended to the floe on the south side, proceeding round the western extremity of the island by my fox-trap, and did not get on board till 2.15 p.m.

On the following days, for four mornings in succession, I again each time witnessed the sun's rise from the top of Beechey Island; the last time the whole disk was plainly visible and distinct above the horizon. During the remainder of the month, and until near the middle of the next, little worth recording transpired.

Bold headland, south side Beechey Island, with fox-trap. (*See page* 63.)

CHAPTER VII.

Lost in the snowdrift—Excursion to Caswall's Tower—Our frozen sherry—Losing our way—The dogs give in—I dig a trench in the snow—We lie down and spend the night in it—All right in the morning—Search parties despatched after us.

Saturday, March 12th.—Ptarmigan having been seen on the floe, after breakfast, at 9.30 a.m., I started on an excursion across the snow-clad plain, in the direction of Caswall's Tower, accompanied by my two Esquimaux dogs, Erebus and Terror, with my old and almost constant companion, the double-barrelled gun, which I had with me both in Parry's attempt to reach the North Pole, and Ross's to reach the South one. Meeting accidentally my young friend Toms on the floe, he volunteered to accompany me, and we consequently started together; the morning most promising, a bright sun in a clear, blue sky, with a light air from the east, so mild at the time that, notwithstanding the month of March in this high latitude is very generally considered, and I think justly so, the coldest month in the year, I left the ship in my ordinary light daily dress.

We struck across the floe for the north shore of the bay, after crossing some of Franklin's sledge-tracks, all in the direction of Caswall's Tower, and from their still very strongly-marked and distinct appearance after such a lapse of time, they must have been made when the soil was in a plastic condition for receiving impressions. We continued our course along the wide, open space, or level flat of snow, extending like a bay floe between the ridge of hills on the north and the Cape Riley ridge on the south. During summer it is interspersed with the lakes and a sparse vegetation, but now it is an entire waste of snow.

When we had got within some two miles or so of Caswall's Tower, we came upon the first appearance of shingle ridges, so low as to be only just perceptible above the snow. Here a pair of the ptarmigan caught my eye, standing on one of these ridges, with outstretched necks of ivory white, and the black row of feathers in the tail strikingly conspicuous in contrast with the whiteness of their plumage and still whiter snow. My dogs, however, alarmed them, and Terror, so unlike her staid brother Erebus, never under control, ran towards them, heedless of my call to check her; consequently both birds took wing when at seventy or eighty yards distance from us, skimming along the surface of the ground, a slight unevenness of which prevented us from marking them down. On reaching the sloping ridge at the base of Caswall's Tower, we came upon some of their droppings and the leaf-buds of small plants plucked off by them, lying on the snow.

This tower, named by Parry, forms a remarkable eminence above Gascoigne Inlet and Radstock Bay, rising above the heads of both to an altitude of some 300 feet; inaccessible on all sides but the north, and even on that side most difficult of ascent, more especially at this season of the year, the bands of snow on its acclivity being frozen so hard as to resist all attempts at hammering out steps with the butt-end of my gun; but just as Toms and I had succeeded in climbing up about one-third of the ascent, a fog bank appeared rising in the east, threatening soon to overwhelm us, and compelling us at once to relinquish all further attempts, although, had the weather continued favourable, I am satisfied we could have reached the summit. My dogs, who were in advance of us, had disappeared for some minutes, and had doubtless succeeded, and were rambling round the top when I made their recall.

At the base we rested for a few minutes on a large, square block of limestone, to feed the dogs, having brought with me a paper bag full of scraps in my pocket for their dinner. My companion took from his pocket a small flask of sherry, frozen, notwithstanding the warm position of his breast-pocket, to the consistence of thick honey, too solid to partake of it in any other way than as a lozenge, after breaking the bottle to come at it. A tongue sandwich he also produced was frozen as hard as the rock we were seated upon, and had about as much flavour in the mouth. This slight and hasty repast being finished, we lost no time in commencing our retreat for the ship, which I estimated to be at least ten miles distant. Unfortunately, I had neither compass nor watch with me—my watch, as I have already related, had been damaged, and as my compass was large enough to be cumbersome, I thought I should not need it to-day, being so well acquainted with the locality.

We had not, however, proceeded far on our return

journey, when the fog became so thick that we could no longer distinguish one portion of the land from another, and after rambling for some time as to our course, a dark, bluff headland peered very indistinctly through the mist on our left. Toms took it for Cape Riley, but no such luck for us, for I well knew we could not have proceeded to anything like the distance that could have brought that cape in sight. Besides, another bluff, dark headland appeared through the fog on the right, so unlike any land opposite to Cape Riley. I could therefore only come to the conclusion that, in the absence of anything to guide our wandering steps, we had got out of our proper course, and been making one of those circles so commonly known to occur to those bewildered in the intricacies of a wood, without any known object to guide them.

It now flashed upon me that we had really gone off at right angles to our course into Gascoigne Inlet, for, from my own long experience in icy regions, I became aware that the sensation to my feet at every step I set was not that produced by *terra firma* beneath them, but from a substratum of water beneath the surface of the hard, frozen snow in which I was treading, and I was soon convinced of this by coming upon cracks and fissures in the floe occasioned by unequal pressure and expansion, which at once removed all doubts. In fact, we were crossing the floe of the inlet, and the insulated mass of black rock looming in the fog on our left was no other than Caswall's Tower itself, and the projecting bluff on the right, the western headland of Gascoigne Inlet.

Under these circumstances, I resolved to take a fresh departure from Caswall's Tower, and we retraced our steps to the very slab of stone at its base on which we had so recently been seated. From this spot I took a careful survey of the bearings of the land around, as a temporary lifting of the fog permitted me, when we once more recommenced our homeward route. For safety, I thought at first of gaining the north shore and keeping close to the base of the range of hills till we were opposite to the ship, but as this would have added at least two or three miles to the distance, with night fast closing in upon us, I determined to keep the middle course of the wide snow-clad plain, as I could still, although very indistinctly, keep the land in sight on either side.

Near the zenith, the fog somewhat thinning off, the Great Bear and the Pole Star becoming visible, encouraged me to continue onwards, shaping my course by the Star of the Pole, which had a position overhead on my right shoulder. This remarkable star, in the constellation of the Little Bear, had for me so many interesting associations of the past. When in Parry's attempt to reach the Pole, the crew at the capstan-bars used to keep time as they moved round in a circle on board the *Hecla* in weighing anchor, to the chorus of—

> "See the North Star glimmering, boys, time for us to go;
> Time for us to go, boys; time for us to go."

This bright star of the second magnitude, only now situated about a degree from the Pole of the heavens, has not always been the Pole Star, neither will it always continue to be so, owing to the precession of the equinoxes, a conical movement of the axis of that Pole, the equinox undergoing a retrograde motion by which it accomplishes an entire revolution in the terrestrial orbit, in round numbers in about 26,000 years, always returning to its initial position; so that in some 12,000 years hence, that fine brilliant star, Vega, of the first magnitude, in the constellation of Lyra, will eventually become the future Pole Star. And another bright and beautiful star, also of the first magnitude, Canopus, will become the South Pole Star, in the opposite hemisphere. Whilst the Pole Star was visible, guided by it, we could shape our course correctly enough for the ship, but before we had got halfway across this desolate wilderness of snow, perhaps some four miles or so from Caswall's Tower, the fog had enveloped all around, completely shutting out from our view land and star alike, and so suddenly that I had barely time to throw my gun down with its muzzle pointing in the direction of the ship, so to get a rough bearing of our position, ere the friendly star disappeared entirely from our sight, leaving us in a bewildering gloom of fog and darkness intermingled, in a death-like silence, with not so much as a rock or hummock of ice to shelter us from the piercing blast and blinding snow-drift.

My two poor dogs had already given in, by repeatedly falling down across each other in a stupor and torpor, from which I had every now and then to rouse them, and drag them by the nape of the neck with sheer force from the spot where they had nestled down in the snowdrift, often having to retrace my steps for some distance in search of them, after every call had been unheeded; my worthy companion, Toms, standing in the same position in which I left him till my return to the spot with the dogs, guided to it by his voice as I dragged both the helpless creatures a dead weight back.

However, he too began, it was too evident, to feel the combined effects of extreme cold, fatigue, and toil telling upon him, which his altered voice, hoarse and half-uttered sentences, proclaimed; yet he displayed a spirit and power of endurance scarcely to have been expected in one so young and inexperienced in scenes like this. In the thickest of the fog, I had no small difficulty in getting him to take my arm, to keep us together.

With the disappearance of the North Star, so disappeared also all our hopes of reaching the ship. The only resource left us was to prepare for the tempestuous night brewing up as best we might, and not risk losing a foot of the distance we had already gained with so much toil and labour, and thus hold our own amid the dense fog which enveloped us.

For this purpose, I fixed upon the softest wreath, or *sastrugi*, of snow at hand to cut a trench deep enough to hold the two dogs, my companion, and myself, with a hunting-knife from Toms' waistbelt, which he had providentially brought with him in the event of shooting a bear, and, although we did not actually come across this animal, we were not a little startled when, in the very thickest of the fog, our dogs rushed back towards us in affright at two objects, magnified as they loomed through the dense mist and drifting particles of ice and snow, into two Polar bears, in the imagination of ourselves, no less than the poor dogs'. Just as I was about bringing my gun to my shoulder, and calling upon my companion to prepare his to receive them, a sudden gust of wind making a temporary break in the snowdrift, revealed to us two immense boulders of rock which, in the excitement of the moment, we had transposed into the shaggy-coated rover and monarch of these wild wastes of snow. But with this hunting-knife, which has led me into this digression, I dug out to the depth of somewhat more than a foot, the only available bivouac in such a fearful night as this, and which possibly might prove our tomb, aided by Toms, who held back the dogs, for these poor animals were so anxious to get into their quarters for the night, that I had no small difficulty in avoiding cutting them during my work. We having piled up the snow around the margin of the cutting in readiness for drawing over us as a coverlet after we had laid ourselves down in it, placing Terror across the chest of my companion, and Erebus at my own feet, we thus ensconced ourselves, at about the hour of midnight, from the direct influence of the war of the elements around, leaving only breathing space kept open by the gauntleted hand in our snow-clad covering. The thermometer was at 32° below zero, or 64° below the freezing-point, Fahr. I cautioned my companion not to yield to sleep, for to sleep under such circumstances would be never to awake. Some of the terrific squalls and gusts of wind that passed over us during the night frequently induced me to get up, and after taking a hasty glance of our surroundings turn in again, quite unable to bear the exposure for more than a few minutes, deepening the trench with the knife as I did so, thus burrowing deeper in the snow as the gale increased.

In this rather novel sort of Arctic bivouac we remained, as near as I could estimate, for some four hours and upwards, when, the fog lifting, and with the morning's dawn the stars appearing overhead to shape our course by, and the land on either side beginning to show, we once more resumed our journey, the dogs and ourselves somewhat refreshed, they by a sound sleep, ourselves from the mere rest.

After proceeding for about three or four miles, we got sight of the ship's mast-head appearing just over the shingle ridge, as we drew nearer to Erebus and Terror Bay. We had long before this seen Beechey Island in front, and Caswall's Tower behind us. From the latter I brought away two rock-specimens as mementoes of this memorable excursion, rendered doubly so to myself, associated as it is, with an escape from shipwreck in the Antarctic seas, when the *Erebus* came in collision with the *Terror* off a chain of bergs in a gale of wind, of which this night is the anniversary of eleven years ago—a somewhat remarkable coincidence, certainly.

When we had got about halfway across the bay, we noticed a party leaving the ship, advancing towards us, which on their coming up proved to be a searching party for us, in charge of old Harvey, our boatswain's mate. We got on board between six and seven a.m., in advance of the party sent out to meet us. The sun at the time was just rising over Caswall's Tower, with a keen head-wind from the westward, having suddenly veered round from the eastward yesterday during the fog, so that we had not even the wind to guide us, although suspecting at the time that such had occurred. I had ventured on the bare supposition to head it for some distance during the fog, and fortunately so far had been right in my conjecture, and made my course good. On getting on board this morning, after a wash and entire change of things, we had a cup of that most refreshing beverage, tea, with a rasher of bacon, in my own cabin.

Sunday, 13*th*.—We both attended divisions and divine service. The searching parties sent out late last evening had not, however, yet left their beds, having been thoroughly knocked up. We both met with the most hearty congratulations from our young friend, Alston, the mate, and the whole ship's company, not one of whom had slept last night, it appears.

After our return on board, we learnt that the whole ship's company had, much to their credit, volunteered early in the evening to form searching parties, under the charge of the petty officers, but it was not till nine p.m. they were permitted to start on their humane errand, the commander expressing his belief that in such weather we must both have perished. However, our generous-minded young friend, Alston, at last prevailed on him to let him have charge of a rescue party; but it appears that, after remaining out from nine p.m. to five a.m., they got so bewildered in the fog, that they found themselves at break of day somewhere under the northern ridges of the bay, opposite to, and in sight of the ship, where they raised a wall of snow, under the lee of which they took up their quarters, but were fortunately all well prepared in warm clothing, somewhat to resist the inclemency of the night.

CHAPTER VIII.

Weather worse than ever—Examining the *Mary* for our proposed spring expedition—Narrow escape from a fatal accident—Laid up with broken ribs—Visit from one of the mates of the *Resolute*—Bird specimens—My first Arctic hare—An Arctic butterfly—Rambling, sketching, shooting—Geological specimens.

Monday, 14*th*.—This has been one of the most severe days of weather that we have yet experienced this winter. The weather, bad as it was during our absence from the ship, since our return on board has set in still worse, yesterday ending in a dismal night, blowing a very hurricane, with drifting snow, and the fog so thick as to entirely conceal the surrounding land, even to the nearest point of Beechey Island, and to-day the only visible object was the dogs' snow-kennel alongside, in which the poor animals lay closely housed throughout the day, all around being a dense atmosphere of overwhelming, drifting snow, through which not a speck of land was visible.

Tuesday, 15*th*.—No change whatever in the state of the weather; blowing as hard as ever, with the temperature as low; all around enveloped in dense fog and mist. How providential was our almost miraculous escape from our perilous position at the moment we effected it. It would have been to the last degree improbable that we could ever have succeeded in reaching the ship, or existed through such fearful weather as we have had since our return, and up to the present time. There is no mistaking the hand of Providence in such a crisis.

Wednesday, 16*th*.—The weather clearing up; fine to-day, and the thermometer even rising above zero. The crew were employed in laying out gravel, where a line had been made for cutting a canal, when the commander, accompanied by the master, his brother, and quarter-masters, went over to the *Mary*, for the purpose of deciding the time she could be got afloat; and as the time for my spring exploration in her is drawing near, I have too good reason to know that any obstacle in the way of my getting her will not be lost sight of.

On each being asked individually his opinion, for fear of giving offence, and knowing the commander's wishes, the quartermaster's reply was not till August or September. Rands alone, the coxswain of my own boat, a thorough seaman and fine, enterprising fellow, with heart and soul in the search, honestly and fearlessly gave his unbiassed opinion that he believed that she could be got afloat in June or July. This candid answer, as may be supposed, gave great umbrage, and so matters stand for the present. In the evening the weather again became overcast and thick, as bad as ever, with immense cracks opening in the floe all round the ship.

Thursday, 17*th*.—A fearful night again, blowing, if possible, harder than ever. I heard the loud howling and roaring of the blast in concert with the harsh cracking and grinding of the floes alongside throughout the night, the whole of the bay this morning presenting a smooth surface from the heavy snowdrift spread over it. Towards evening the gale again lulled, and I could see Caswall's Tower from the ship; but in the night it came on to blow again, with thick weather.

Friday, April 22*nd*.—After breakfast this morning, on going down to the lower-deck to see a patient slung in a cot there, upon crossing from one side of the deck to the other, the hatch having been most carelessly left off, I was very nearly being precipitated to the bottom of the hold as, in the dark, I stepped over the edge, and only saved myself by grasping in my descent the "combings" of the hatchway with one arm, and when so suspended, by a desperate effort threw my body above the hatchway. The only light was a "dip" burning by the sick man's hammock. I came off with a fractured rib or two, and laceration of the intercostal muscular fibres, confining me to my cabin for about a fortnight for the bones to unite, under my young friend Toms' care and attention, which was unremitting. The pain in the right side, which came in contact with the sharp edge of the hatchway in my fall, was for some time afterwards very acute. My time during the preceding month was chiefly employed in my cabin, finishing my charts, sketches of land, and writing up my journal and reading.

Saturday, May 14th.—The sun since the commencement of the present month has appeared above the horizon throughout the whole twenty-four hours. On the 30th, about noon, as I was occupied in arranging papers, books, &c., in the drawers of my cabin, I heard a report on the main-deck that a stranger was seen coming over the shingle ridge from Union Bay, and making for the ship, and upon going on deck was not a little surprised to see Roche, one of the mates of the *Resolute*, coming up the ship's side alone, and on shaking hands with him on deck, learnt, to our astonishment, that the *Investigator* was at Bank's Land, having made the Polar Passage. Another officer and man arrived soon after, and Alston introduced me to Wynniatt, one of the lieutenants of the *Investigator*.

Tuesday, 31st.—Blowing half a gale of wind, with raw, foggy, overcast weather. At four p.m., Alston, with the dog-sledge and three men, left the ship to meet Lieutenant Cresswell and fourteen men sent from the *Resolute*. I had a long chat with Wynniatt and Roche, from whom I learnt that good old Captain Kellett had desired to be kindly remembered both to myself and Toms.

Thursday, June 2nd.—I was awakened at four o'clock this morning by the arrival of Lieutenant Cresswell and party. On the 4th, Roche, accompanied by one man, started with the dog-sledge to rejoin the *Resolute*.

Tuesday, 21st.—The longest day; cloudy, but mild and calm. I walked round Beechey Island for the first time during the last three months, and found it very different now to what it was in my winter excursions. Having started at ten a.m., by the Union Bay ridge, I did not get on board again till two p.m. The snow covering the floe had become so soft, that in many places I sank above the knees at every step; some spots were sludgy, with water beneath, filling my fisherman's boots over the tops with snow and sludge, rendering the entire journey a most laborious and toilsome one. I picked up the first specimen of the *Saxifraga oppositifolia* that I have seen in bloom this season, and collected a few fossils from the ridge.

To-day I began a series of sketches for a complete panorama of Beechey Island all round, commencing my subject with the lofty, precipitous crag from which the Union Bay spit runs out, near my fox-trap. Here I got a glance of the raven as he flew croaking to the rocks, on which he alighted, but took especial care, prudent bird, to keep out of range of my gun. Saw a dovekie or two, and a small colony of the glaucous on the eastern precipice, where they are now breeding, uttering their wild, harsh cry of "Qua, qua," somewhat resembling the quack of the duck, at other times a more euphonious note. One fine bird flying overhead, a tempting shot, but towering far too high for small shot to have any effect, I fired the barrel loaded with ball, the first time I have fired it off for some three months past, and had the satisfaction to see my bird come down with a rotatory motion, and with such impetus as to bury itself in the snow at my feet, the aim being almost vertical. It was a fine male, measuring five feet from tip to tip of wings, and weighing four pounds. I preserved the skin, which I still have in my collection; the ball went right through the body. Toms and I had the body for dinner.

Friday, 24th.—Midsummer Day; somewhat resembled a Christmas one in England. At eleven a.m., accompanied by my two dogs, I strolled beyond the *Mary*, and completed my series of sketches of Beechey Island, with one of the glaucous, and the other of the raven rookery. Numbers of the glaucous were sitting upon the rocks and the ledges containing their eggs, hopelessly inaccessible, as were the ravens also at home, croaking loudly all the time I was making my sketch.

Monday, July 11th.—The commander, having started this evening with a sledge party for Cape Beecher, was about pressing into his service two of my boat's crew, Millikin and Nugent, whilst they were at work on the ice, but I at once checkmated this poor attempt to deprive me of a boat's crew by directing Toms to place them both on the sick-list, with my own servant, Clark, the marine.

Wednesday, 20th.—I made an excursion round Cape Spencer for some three miles beyond, as far as a lake and large watercourse, where I rested, and lunched on some preserved beef and biscuit, and a bottle of lime-juice, leaving the empty bottle as a memento on a pile of limestone rock overhanging the steepest part of the rapid current, as it foamed along its narrow channel to empty itself into the Wellington Channel, where a long ledge of rocks enclose a lagoon, in which some heavy masses of ice have been confusedly piled to a considerable height, looking yellow, decayed, and old. Got on board at six p.m., having left the ship at ten a.m.

Thursday, 21st. After breakfast I started on a geological and botanical excursion to Cape Riley, about four miles distant across the bay floe. A crack in the ice leaving some open water, was full of dovekies, and extending from the cape. On landing, I left everything but my gun near the Beacon Cask, and began the ascent of the cape at a rather difficult spot. I found the fossil shells tolerably thickly distributed between the base and the summit, but for the most part much weathered and decayed from atmospheric exposure during the lapse of ages. The productæ and corallines were most numerous, the former nearest the summit. This shell, the producta, first appears in the upper Silurian, and continues throughout the Devonian and carboniferous periods. The day, being one of the most lovely ones—with a brilliant sun—we have yet experienced, a clear, blue sky, scarcely flecked by the faintest cloud, with a light air from the eastward, afforded me a magnificent prospect of the surrounding lands, with Barrow's Strait and the Wellington Channel extending white and level as far as the eye could reach in one unbroken, vast plain of ice.

After a somewhat difficult descent, with my pockets filled with fossils, some few collected from the summit, I saw a white object spinning round and erect, like a phantom, amongst the rocky ledges at the base of the cape, amongst which it suddenly disappeared, but not before I had recognized it to be the Arctic hare. Following this beautiful creature along the *débris* where I had lost sight of her, on looking down upon the plain beneath I again caught sight of her seated there. Descending from the ridge, I attempted to get within shot, but white puss was too wary for that, and bolted off, followed by a ball from one of my barrels when about 100 yards off. Following up my game with the remaining barrel, loaded with No. 4 shot, when I got within about fifty yards of her I fired, and shot her through the head, and she rolled down the ridge in convulsive struggles, bleeding profusely from the nostrils, to the injury of the beautifully white fur. This is the first hare we have met with, and the first I ever shot; a full-grown animal, tips of the ears black.

Here I caught a small brown butterfly, and about half a dozen very small dipteræ dancing in the air, only a few feet above the surface of the ground, like the ephemeræ towards the close of day in England. How these frail and delicately-organized creatures can spring into existence here is really most wonderful. I saw a solitary snow-bunting and the raven, with now and then a gull flying over the point, and before I left I shot a fine silvery gull. I had some preserved beef and a slice of bread, washed down by a draught of pure water from an adjacent running stream, for my dinner, on the point of Cape Riley. I started on my return, with as much as I could well carry, haversack full of fossils and plants, and did not get on board till ten p.m. Toms came on the floe to meet and congratulate me on my success. When I had finished laying out my plants in papers, it was midnight before I turned in, the sun shining into my cabin port.

Friday, 22nd.—Being the anniversary of my fifty-third birthday, I passed the day in an excursion to Cape Riley. The weather was cloudy, with a fresh north-west wind, but not so fine as yesterday. On reaching the point extending out from the cape, I was fortunate enough to discover two more hares quietly squatted under the lower ridge of rocks, one somewhat higher up than the other. I succeeded in shooting them both, a male and female, and I found them no small weight to carry for above three miles over the floe, getting on board about three p.m. I shot an ivory gull (*Larus eburneus*) near the ship. The crew were working at the triangles on the ice, and looked somewhat astonished at seeing me freighted with two more hares. I took a sketch of the bay and surrounding hills from Cape Riley. As soon as I got on board, I had my lime-juice draught, that excellent antiscorbutic. One of my hares weighed nine pounds, about the weight of the Arctic fox.

Tuesday, 26*th*.—Cloudy and cold, with south-east winds. At nine a.m, I crossed over the bay floe to Cape Riley again, but on finding no more hares there, I followed the line of beach under the cliffs to the top of the bay, and on reaching the plain leading to Caswall's Tower, I had a shot at four eider ducks, firing both barrels, but without effect, they being beyond the range of my gun. After shooting a sandpiper, I ascended one of the cliffs in the northern range of hills, having a lake at its entrance.

From a broad-bottomed, deep, shingly ravine, having a rapid watercourse down its centre, I ascended a high hill by the watercourse. Here, amongst the broken-up *débris* of limestone, having a foliaceous fracture, I found the largest and most perfect specimens of madrepores, and corals with echini, profusely scattered about the acclivity of the hill, with some fragments of ironstone, of which I collected some fine specimens, and also of the cup coral (*Caryophylia cyathus*), a fossil of the upper Silurian series.

This hillside, richer both in the profusion and perfection of its specimens of the vestiges of past ages than any other locality which I have hitherto come across here, impresses the mind with the conviction that creatures constructed as these are could only exist in the warmest regions, therefore must have at some period or other lived in a tropical, or at least subtropical climate here. For the very perfect condition of the delicate organization these specimens are found in, precludes all supposition as to their having been transported from any distance.

When on this limestone table-land of North Devon, I had to make a circuit of the heads of the ravines, which terminate in precipitous cliffs for a mile or more inland of the bay, filled with snow or rivulets. This was harassing work, adding several miles to the toilsome route before I could regain the plain beneath. This I at last accomplished by descending a ravine, following the watercourse till it passed through a tunnel in the ice and snow, when I ascended the bank on the right, and finally descended to the plain. Gathered some fine specimens of that elegant golden yellow *saxifrage* (*Saxifraga flagelaris*) with its curling tendrils, like strawberry runners or spider's legs in bloom, growing on the mossy margin of the lake. I returned on board about eight p.m., having been eleven hours from the ship, and allowing the moderate rate of two miles an hour, inclusive of stoppages, I must have gone over twenty-two miles of ground.

Monday, August 1st.—Clouds and sunshine, with a fine westerly breeze. After breakfast, I left the ship on an excursion to Caswall's Tower. Proceeding along the Cape Riley ridge, I shot a leveret on the rocky ridge, near where I shot the hares, and also two ravens flying overhead. Having placed the whole *en cache* beneath some rocks, I ascended to the top of Cape Riley, and just before I reached it, shot a third raven, leaving it also behind me. From this spot I had a fine view of the water lately opened out by the crack between Cape Riley and Beechey Island, the separation in the floe forming a channel two or three miles in width, diverging off at an angle in the direction of Cape Bunny as far as the eye could reach, but the entrance to Wellington Channel remained as closely beset as ever. I had a long and laborious journey along the whole length of the ridge, having to cross some half-dozen of deep, flat-bottomed ravines, with watercourses running through the centre of them, here and there a large patch of snow remaining.

The shingle table-land was more easily passed over, especially approaching Caswall's Tower, at its eastern extremity. After travelling some six miles over the summit, Gascoigne Inlet appeared below it, into which I descended down a somewhat steep, shingly hill, and across a large patch of snow at its base, where I heard the lively call of the sandpiper, and on my reaching the plain below, saw the little bird running along on its slender legs by the margin of a stream, and might have shot it, but as I had already obtained the requisite number for the collection, three specimens, I felt reluctant to deprive the pretty little creature of the existence which it appeared in the full enjoyment of, and so happy even amid this scene of desolation, but which was natural to it, and still more, its summer home.

I next crossed the floe over Gascoigne Inlet, which, with the exception of a few cracks on the opposite shores, proved smooth and good travelling for about two miles.

CHAPTER IX.

Successful ascent of Caswall's Tower—Traces of an old Esquimaux encampment—Gain the summit—Collection of specimens—The *Phœnix* and *Breadalbane* come in sight—Lieutenant Bellot—Commander Inglefield—Bellot's undertaking—I am appointed "additional" to the *Phœnix*—Bellot's death reported by his survivors—Loss of the *Breadalbane* in an ice nip—I get my baggage safely on board the *Phœnix*—The *North Star* almost deserted.

THE last attempt to ascend Caswall's Tower baffled our efforts. I was still at least a mile off, having a shingle bank between me and it. On reaching its base, I saw the remains of very old Esquimaux encampments, forming circles of limestone slabs, intermingled with the bones of whales partially embedded in a rich bed of mossy turf; the ribs, pelvis, and vertebræ bore the marks of great age. These spots were much more richly clothed with mosses, lichens, and flowering plants, and the flowers larger; the most striking blooms were those of the poppy, the saxifrages, both the purple and the golden flagelaris, with its slender purple tendrils, the yellow ranunculus and campanula, and a pretty little yellow flower much resembling the golden saxifrage. Doubtless, these fine, large blooms owed their vigorous appearance and unusual size and colour to the richness of the soil, formed from the decay of whales' blubber and other animal remains at the time it was the home of the Esquimaux.

The circles of stones measured about twelve paces in diameter; the bones were stuck into the ground between the slabs of limestone, and from the number of holes in

R. McCormick, R.N., del. Vincent Brooks, Day & Son, Lith.

View from the summit of Caswall's Tower, 300 feet. The foreground was the scene of our night's bivouac, my conpanion, and Esquimaux dogs, Erebus and Terror, in the snow, with thermometer 32° below zero. Erebus and Terror Bay, Beechy Island, and the Wellington Channel in the distance.

PAGE 88—VOL. II.

the vicinity, both hares and lemmings find a refuge here. The droppings of the former were thickly scattered about, although none appeared to be very recent. I had a distant shot at two Arctic gulls and an eider duck.

I ascended Caswall's Tower from the west side, the only side on which it is accessible—and this is very steep, in some places almost perpendicular, over piles of foliaceous limestone, which somewhat lessens the difficult, although laborious work. It is between 300 and 400 feet in height, having a flat, table-topped summit, covered with dry mud, intermingled here and there with limestone fragments, and scattered over very sparsely with a few small, stunted poppies and saxifrages. On walking round the top, I found it bounded in every direction but one by fearful, precipitous escarpments, wild and picturesque crags, and pinnacles of rock, rising from below in chaotic confusion, almost producing a sensation of giddiness in looking down upon them.

As I was piling up a small cairn on the south side, near to a craggy, projecting point, under which I placed a ball with a slip of paper, on which I had written my own name and that of the ship in pencil, I most unexpectedly came upon a lemming on raising a small slab of the limestone, beneath which a young lemming had concealed itself, flat as a pancake. I caught it, and took it on board with me, and still have it preserved in my collection. Three or four white balls of hare's fur, and resembling the castings of a bird of prey, were lying near. I picked up a few plants, and the fossil *Tubipora*, and sketched the entrance to Gascoigne's Inlet, Beechey Island, Cornwallis Land, &c.; and after having my lunch of preserved beef and bread on the summit, I accomplished the descent only a few paces to the north of that by which I ascended to the shingle ridge on the west side to the Esquimaux ruins.

At the head of Gascoigne's Inlet I saw four red-throated divers on a lake. As I returned along the plain, I shot a solitary king-duck, flying past me at the distance of some fifty yards or so—the only one of the species that I have been able to obtain. I had to cross several deep watercourses, and to walk for some distance along the head of the bay, before I could get upon the floe, in consequence of its having separated from the shore, leaving open water. It has now become excellent walking upon the floe-ice, the pools of water having drained off, and the slippery ice become covered with a crisp coating of snow, of a smooth and uniform whiteness. The pools were this morning frozen over. On nearing the ship, I had to make a long *détour* to avoid the docks and cuttings. My two dogs were lying alongside, but nobody on deck.

Thursday, 4th.—After breakfast I started across the floe for Cape Riley, to bring on board my *cache* of Monday last. Although three days and nights had elapsed, I found both leverets and ravens quite dry and uninjured, yet we had had almost continuous rain. I saw another leveret on the ledge of rocks, and on approaching it a second appeared above it, standing erect on its hind legs, looking at me, and spinning round in the same fantastic way as the first I shot here did. I shot both of these, and then ascended to the summit of the cape to pick up the third raven, left there on my last visit. The mountain was now suddenly enveloped in a dense fog for a few minutes, which lifted before I reached the base. I picked up a specimen or two of fossil shells, weighing about five pounds, which I brought on board in my pocket, distributing the three hares and three ravens about my person as well as I could, slinging the latter to my shot-belt, which formed a girdle round my waist, with one hare in my haversack, another suspended from the barrel of my gun, carrying the third in my hand. The whole formed a good freight, considering the nature of the floe across which my course lay. As I passed the triangle at which our crew were employed cutting with the saw a canal through the ice, they looked astonished at my Robinson-Crusoe-like appearance, so laden with game from so desolate a region. I got on board at three p.m. Barrow Strait seemed to be much more open, and the ice in motion, drifting fast down channel.

Sunday, 7th.—Nugent, one of my own boat's crew, brought on board for me a young lemming, found by my dog Terror under a stone, just below the ravine of Beechey Island. In the evening, it ate readily and heartily of some willows and other plants Toms had kindly collected for it. It frisked about in a lively manner, uttering a sort of squeaking bark. I put it into a box in a nest of tow, into which it soon rolled itself up and slept.

Monday, 8th.—Cloudy, with westerly wind. Between three and four p.m., two vessels hove in sight, coming round Cape Riley, announced by a "hurrah!" from the floe. They proved to be the *Phœnix* steamer, commanded by Captain Inglefield, and the *Breadalbane*, No. 2 transport, with provisions and coals for the squadron. I happened to be walking the deck when Commander Inglefield came on board, who, raising his hat to salute the quarter-deck, said "How do you do?" This was all that then passed between us, as we had never been introduced to each other, and he descended at once to the commander's cabin. Charlton, the surgeon of the *Phœnix*, whom I had previously known at Woolwich, came to see me, accompanied by the young Frenchman, Lieutenant Bellot, with whom I had a long chat, several other officers of the *Phœnix* being with them.

Tuesday, 9th.—We have to-day completed an entire year in this bay. The ships were lying off Cape Riley, unloading their stores and coals there. After breakfast a party of us walked across the bay floe to them. On going on board the *Phœnix*, Dr. Charlton introduced me to

Commander Inglefield on the quarter-deck, and I had some conversation with him about his recent voyage up Smith's Sound, when he remarked, "I very anxiously looked out for you there." I told him that I was not permitted to start until the short season was so far advanced winter was setting in, and all that I had time left me to do was to set at rest the mooted question as to whether any opening or means of communication existed between Baring Bay and Jones Sound.

I told him of my intention to return to England with him, as I could be of no possible further use out here, circumstanced as I was, precluded from doing anything further in the search. He replied that he had instructtions from the Admiralty to bring home all who wished to return. We all dined on board the *Phœnix* at two p.m., and I had a long and interesting conversation with Bellot on deck, and learnt from him at the dinner-table that my lamented dear friend, Lieutenant Le Jeune, of the French navy, was dead, who had been the companion of Duperrey in his voyage round the world. In the evening, Bellot and Osborne pulled us in the whale-boat along the floe-edge part of the way back to our ship, and from this we struck across the floe through a dense fog, entirely concealing the surrounding land. Saw several large seals in the water. We got on board about nine p.m., after splashing through numerous watercourses and pools of water on the ice, at times over the tops of our fisherman's boots, and jumping over others.

Friday, 12th.—Calm, cloudy, overcast day. Bellot this morning breakfasted with us in the gun-room. He is to start this evening with despatches up the Wellington Channel. He came into my cabin to see my sketches of the coast-line, and copied some of the headlands into his note-book. He read the letter to me from Captain Hamilton, the secretary to the Admiralty, and also the one from my friend, Mr. John Barrow, and said that he had met the latter at Sir Francis Beaufort's, with the Admiral of the Fleet. At this moment I was sent for on deck, and a memorandum put into my hand, directing me to transmit to the commodore my track chart and coast-line sketches, which I set about at once, accompanying them by a letter "on service."

Bellot came on board this evening in his travelling-dress, most picturesque in appearance, as he walked our quarter-deck. It consisted of a light salmon-coloured robe, or frock, having a hood, and belted round the waist with a leather girdle, from which appeared a hunting-knife in a leather sheath; fisherman's boots, above the tops of which the white hose showed; a Welsh wig, with no other covering to his head: he held a staff, having a square piece of wood at the end, in his hand; a small bag was suspended from the girdle, containing the "despatch" for the commodore. Thus equipped, this gallant young fellow started, accompanied only by four men from our ship, a sledge, and a "Halkett's" indiarubber boat, on a journey of so hazardous a nature, amongst drifting ice, full of broken-up floes, acted upon by winds and currents, in such a dangerous navigation as the Wellington Channel is, especially at this season of the year.

On no account whatever should he have been allowed to depart without a whale-boat, or some other equally efficient one. This precaution, so necessary, as I had found from my own personal experience, I endeavoured to impress upon him, as a paramount duty he owed both to his followers and himself, and that if a boat was not offered him by the commander of the *North Star*, he should himself apply for one. Could this noble and enthusiastic, unassuming youth have been induced to do so, it could not have been refused him under such circumstances, and we now might not have had to deplore the premature loss of one of the greatest ornaments of the navy of France, or, indeed, of any other country.

Saturday, 13th.—I caught my little lemming on the deck of my cabin this morning before breakfast, and returned him to his box, which he by no means approved of, making angry efforts to get out by gnawing the wooden bars, and chattering with his teeth, so I let him out again, limiting his excursions to the space in front of my port-sash during the day, which delighted him so much that he gambolled round me, running under the book before me in the most playful manner.

Saturday, 20th.—Yesterday having been appointed "additional" to the *Phœnix* for a passage home, I joined her to-day, walking over the floe to Cape Riley, where she remained beset in the ice. I was accompanied by a party and most of the *North Star's* ship's company, all, indeed, that could possibly do so had volunteered for the *Phœnix*, so thoroughly dissatisfied and sick of the ship had they become during the wretched, miserable winter passed in her.

To-day two of our men—Harvey, the boatswain's mate, and Madden, a seaman—who had accompanied poor Bellot, returned with the melancholy intelligence of his death. He was blown off a hummock of ice in the gale up the Wellington Channel on Wednesday last, when drifted from the shore in the turmoil of the pack-ice, between four and five p.m. I learnt from Captain Inglefield that poor Bellot had left a sister in ill-health dependent on him for support.

Sunday, 21st.—I was awakened this morning about four o'clock with the report that the *Phœnix* was in a "nip," and the *Breadalbane* transport, just astern of her, in a still more severe one. Upon going on deck, I saw the transport heeling over amid the ice, about midway between Cape Riley and Beechey Island.

I witnessed the whole catastrophe from the stern of the *Phœnix*. It occurred during a dead calm; the horizon in the south dark, having a diffused blackness. The enormous pressure of the contending floes was such as to crush in her bottom, sinking beneath the overwhelming ice within a quarter of an hour from the commencement of the "nip" to the disappearance of her top-gallant-masts beneath the floe, which closed over her, leaving scant time for her crew to jump upon the ice, with their bags and hammocks, to save their lives.

Monday, 22nd.—A fine day, and the ice, fortunately for me, opening, enabled a boat to bring away my things from the *North Star*, for which purpose, after breakfast, I walked over the floe. By prompt action, hurrying my baggage and my two Esquimaux dogs into the boat alongside, we made our way, without the loss of a moment, through the temporary canal formed by the separation of the floe, and which might close up again at any moment, and on reaching the *Phœnix* with my baggage, which I was within an ace of leaving behind me, I got it struck down her spirit-room. The *North Star* is now left without a single A.B. of her own crew, and only five hands remain altogether on board of her. This of itself indeed speaks volumes.

Peaked mountain, Graham Harbour, I ascended, by the dotted track up and down. (*See page* 97.)

CHAPTER X.

On the way home—Weather-bound in Graham Harbour—Disco—Festivities—Anchor at Holsteinberg—The Moravian chapel—Death of my dog Terror—We near the Orkneys—Home.

Wednesday, August 24th.—At three p.m. we took leave of Cape Riley, Beechey Island, and the bay—where I had passed just a year of my life—under steam, with light winds from the westward, and foggy weather. About seven p.m. we passed Radstock Bay, afterwards Cape Hurd, and before midnight we were off Maxwell Bay.

Thursday, 25th.—At 4.30 a.m. our course was arrested by the pack, resting on a point near Hobhouse Creek. A boat was sent away to look for a harbour, when, a thick fog coming on, we eventually got into a creek which was named Graham Harbour, and within which the heavy pack soon closed us in. To-day, I dined with Commander Inglefield in his cabin, at which several officers of the ship were present.

Wednesday, 31st.—We have been weather-bound in an amphitheatre of hills, with a heavy easterly gale blowing, and snow, sleet, and rain at intervals, since the 25th; and this morning, the weather having cleared up fine with sunshine, I landed, and ascended the peak on the eastern side of the entrance to the harbour, 1000 feet in height, to ascertain the condition of the ice in Barrow's Strait, and the chances of our escape from being frozen in here for the winter.

After a toilsome ascent, first over loose blocks of limestone, then up two craggy projections or buttresses, I gained a saddle in the mountain, near the summit, when I had before me a very steep conical peak, ice and snow-clad, rising for some fifty feet to the summit; by beating steps in the hard, frozen snow with the butt-end of my gun, I had approached so near to the perpendicular with such a slippery foothold, that I was, when so near the top, compelled to descend on the side next the ravine, over a surface so precipitous and slippery, that I had to beat out every step for foothold on the frozen surface with my gun.

I was rewarded, however, by a fine commanding view of Barrow's Strait, and from the highest point attained I saw two extensive lanes of open water, as far as the eye could reach from ten to fifteen miles off shore. And of this we were immediately enabled to take advantage, and at once got out of the creek, finally named Phœnix Harbour, after the ship. When I reached the bottom, I found the surgeon and several other officers of the ship awaiting me there, and we returned on board at 3.30 p.m., when, as soon as dinner was over, we weighed anchor, and steamed out, but had to anchor again at ten p.m. Whilst detained here, I took sketches of the harbour all round.

Thursday, September 1st.—Being finally out of the creek, we steamed along a narrow lane of water, between the pack and the shore, forcing our way through some necks of ice, and heavy masses. On the following day, when off Port Dundas, in Croker Bay, three native canoes came alongside from an encampment on shore, where five tents and a small group of the Esquimaux were visible. In the afternoon we got clear of Lancaster Sound, and on the following morning found ourselves off Pond's Bay, with a favourable wind for Disco.

Friday, 9th.—We anchored in the harbour of Lievely, and remained there until Saturday, 17th, where we passed a very agreeable week.

Saturday, 10th.—I went on shore accompanied by the surgeon of the *Phœnix*, and Lieutenants Wynniatt and Cresswell, in a native boat, visited the native hut, where we found the governor's two attendants, the sisters Sophie and Marie, and afterwards called on Mr. Olick, the Inspector-General of North Greenland, by whom we were most hospitably received, and wine placed before us, brought in by Marie. I learnt from Mr. Olrik that the former governor, Maltrop, was now at Ritenberg, and his clerk at Egistemond. We next visited the governor, a young man, where wine was also set before us by Marie. Afterwards we took a ramble round the hill, and a visit to the churchyard, returning on board to dinner.

Sunday, 11th.—Captain Inglefield dined in the gunroom, and during the forenoon we had a large party of the natives on board to see the ship. The music-box was set going, to which they sang songs, and had some wine and biscuits. Sophie with her mother and two of her sisters were among the number; and on my presenting her with a case of needles, she made signs that she would make me a watch-case. The Esquimaux men of the party had a glass of grog each, and they all left much gratified.

Monday, 12th.—Dined in the cabin with Captain Inglefield to meet the Inspector-General Olrik, and the governor and his lady, the *Investigator's* officers, and some of his own. Mr. Olrik afterwards joined us in the gun-room, and I had a long chat with him. On shaking hands with him, as I saw him over the ship's side with Captain Inglefield, he gave me an invitation to come on shore and sup with him this evening.

The supper-table was laid out at the governor's, and he and his lady presided. There were present besides the Inspector Olrik, Captain Inglefield, three or four of the officers of the *Phœnix*, and myself. Marie attended at table. The governor presented me with a curious toothpick, made out of a walrus tusk, by one of the Esquimaux, very ingeniously, with only a common knife. Two salutes were fired whilst at the governor's, and one of ten guns on Mr. Olrik's coming on board the *Phœnix*. We left at eleven p.m. When I returned on board in an Esquimaux boat, the night was dark and stormy, with a loud crashing among the bergs.

Tuesday, 13th.—We had an invitation from Mr. Olrik to tea with him this evening, and at eight p.m. we landed. An excellent supper, with all the produce of Disco, spread out before us. Small radishes grown in the governor's garden, excellent rye and white bread, with fresh butter, cold glaucous gull or burgomaster, washed down by good claret, &c., and a good cup of tea. Governor present, and Sophie attended.

Wednesday, 14th.—I landed and paid a visit to the Esquimaux huts, constructed of turf and willow, with wooden beams, and windows formed of dried intestines.

At the Brobergs' wigwam, which was superior in its construction to the rest, Sophie and Marie's family consisted of eleven in all—father, mother, and six daughters, with five sons. I met Mr. Olrik, the Inspector-General of North Greenland, who took me home with him, where I had a cup of coffee, and saw the radishes growing in the garden, and brought in by Sophie.

Thursday, 15*th*.—After breakfast I accompanied the surgeon of the *Phœnix* on shore to see a patient, one of the Brobergs' daughters, and a sister of the governor's two attendants, Sophie and Marie. Mr. Olrik himself came in, and acted as our interpreter, when, having duly prescribed for the patient, we returned with Mr. Olrik to his house. He showed us round his garden, the carefully preserved radish-beds, and workshops, and upstairs to his library, where, at my request, Sophie wrote her name on a sheet of paper, as a specimen of her handwriting, and in return I presented her with a new silk kerchief for her head-dress, putting it on for her according to the latest Esquimaux style of fashion, when she instantly shot out of the door laughing and delighted, and much to the amusement of the inspector-general, who also gave me a specimen of Esquimaux composition in the native language. We next visited the school, at which about fifteen little girls and boys were seated at table learning their letters; this had been the chapel. Here Captain Inglefield joined us, and after calling on the lieutenant-governor, we returned on board about noon, through a shower of rain.

The inspector-general and lieutenant-governor were saluted with nine guns on coming over the ship's side to-day, and were received at the gangway by all the officers of the ship. We dined at three p.m. After dinner I had a long chat with both the governors in the engine-room, to which they had retired to smoke their pipes. I presented each of them with copies of the

Arctic blue-books containing my own plans of search, the last new navy list, and the new chart of Baffin's Bay. I learnt from the inspector-general that he had passed five winters and seven summers here, and that the governor had been previously at Egedesmunde.

Friday, 16th.—This morning the first lieutenant of the *Phœnix* delivered to me a very handsome model of an Esquimaux canoe, which had been made expressly for Inspector-General Olrik, who presented it to me as a parting token, and which was brought on board last night by the first lieutenant.

After dinner I went on shore to take leave of Mr. Olrik, and thank him. He showed me his collection of specimens, gave me some sea-birds' eggs, a bottle of mollusca, and a very old volume of "Egede's Greenland Journal," also an introduction to his brother, a merchant in London, with his card of address.

On Sophie coming into the room, I told her, when she had finished the watch-pocket she was making for me, to give it to Mr. Olrik, which he interpreted to her, and took me on board in his own boat, having engaged to take Captain Inglefield on board of the Danish barque, of 320 tons, the *Northern Lights*, lying in the harbour, and we all went on board the barque together, under a salute of nine guns, one on each side, and very small ones too.

Her captain, Bank, had been out here no less than thirty-nine times, and wintered once. He handed us each a glass of champagne, with the Danish custom of touching glasses together in drinking healths. The cabin was very comfortably fitted.

On our return on board, the kind-hearted and hospitable Dane, Olrik, took his final leave of us. From this fine specimen of the Dane all the English expeditions out here have invariably met with the greatest kindness and hospitality.

During the first watch he sent on board the watch-pocket which the kind Esquimaux girl had made for myself, as a specimen of native work, also a black Esquimaux pup, from himself to Captain Inglefield. The cables were shortened in, and boats hoisted up and in, preparatory to getting under weigh at four o'clock to-morrow morning.

Saturday, 17th.—At six a.m., under a salute from the barque and the two small forts on shore, which we returned, we stood out of the harbour of Lievely. Fog, snow falling, and hills white.

Sunday, September 18th, at two p.m.—On anchoring here, Holsteinberg, we found our old whaling friend, Captain Parker, in the *Truelove*, at anchor in the harbour. He, with the two governors, came on board. After dinner I landed for the first time in Holsteinberg. Called on the governor, who, with the lieutenant-governor, occupy the same house—the latter on the ground floor, and the former the first floor.

Jorgen N. Neilson Molher, the governor, so resembles in personal appearance Shakespeare's fat, jovial knight, both in bulk, dress, and manner, that I at one dubbed him Sir John Falstaff. He possessed a fund of merriment, singing capital songs, and treating us with great hospitality to reindeer, ptarmigan, and other delicacies of the country, with no lack of claret, punch, gin, &c.; and from the abundance of animal spirits he displayed throughout the evening, must, I should imagine, be the very life and soul of this little community, shut out as it is from the rest of the world throughout the long, dark, tedious winters.

Monday, 19th.—I went on shore again, and paid a visit to the priest and the lieutenant-governor, Elberg, whose ladies offered refreshments in the shape of the country's preserves (*Empetrum nigrum*) and cherry bounce, and Mr. Elberg afterwards accompanied me as interpreter in the purchase of a large Esquimaux canoe I was in treaty

R. McCormick, R.N., del. Vincent Brooks, Day & Son, Lith.

The Danish Settlement of Holsteinberg, North Greenland.

PAGE 102—VOL. II.

with an Esquimaux for, and purchased, with all its gear, for the sum of 2*l.* I also went with him to the government store, and bought some reindeer skins. Eider-down was 10*s.* per lb. I paid a visit to the little Moravian chapel, which contained, besides the pulpit, an organ, communion table, and something like a score of benches. the whole was surmounted by a belfry. I took a sketch of the harbour and settlement.

Tuesday, 20*th.*—A lovely day, a bright sun in a clear blue sky, the finest day we have had this season. I turned out at five a.m., went on deck, and we weighed at 5.30 a.m., in company with the *Truelove*, towing her out of the harbour, till she cast off from us at four p.m., with three cheers, and before darkness set in was hull down astern.

Friday, 23*rd.*—We passed the latitude of Cape Farewell at 59° 36′, and longitude 51° 35′. We have had fine displays of the aurora.

Monday, 26*th.*—One of my poor dogs, Terror, was taken ill with some virulent distemper, and notwithstanding all my efforts to save her life by medicine and unremitting care, she died. Her skin was in such a condition I could only preserve the carcase in pickle for the skeleton, now in the Museum of the Royal College of Surgeons.

Tuesday, October 4*th.*—This evening we were off the Old Man of Hoy, in the Orkneys, and entered the harbour of Stromness in a gale of wind and rain.

On the following day, Wednesday, the 5th of October, Dr. Hamilton came on board, and I accompanied him on shore to his house, where he introduced me to Mrs. and Miss Hamilton, with whom I chatted till two p.m., when he returned on board with me to dinner, and afterwards I returned on shore with him, and had tea with his amiable family; his wife is a sister of the distinguished Arctic traveller, Dr. Rae. There are nine children. I passed a very pleasant evening, and left about eleven p.m. On the following day, Thursday, the 6th, I went on shore again, and took leave of the family, when, about one p.m., a gun was fired, and the "Blue Peter" hoisted, and soon after I got on board we got under weigh.

Friday, 7th.—My dog Erebus was seized with the distemper, but the experience gained in Terror's case enabled me perhaps to save him, for he recovered. On Wednesday, the 12th, we saw the lights of Cromer, Hasborough, Winterton, Newarp, and Cockle Gat, and passing outside the sands of Yarmouth Roads, we arrived at Woolwich on Friday, the 14th.

The remarkable nature-sculptured mountain-limestone hill at the end of this chapter is a facsimile of the mountain, had the engraver rendered the parts in relief more rugged in outline and less artificial. It is situated near the entrance to Lancaster Sound.

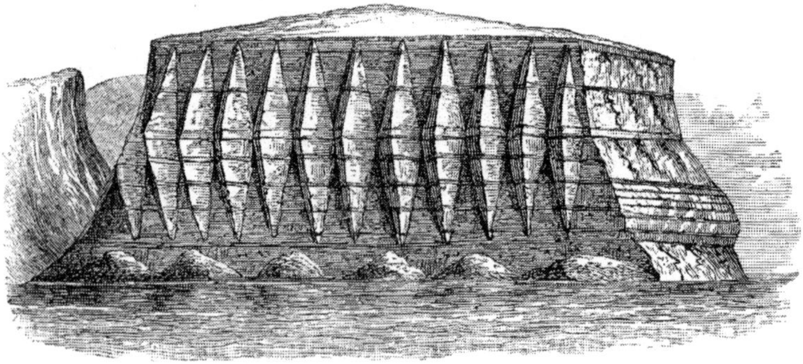

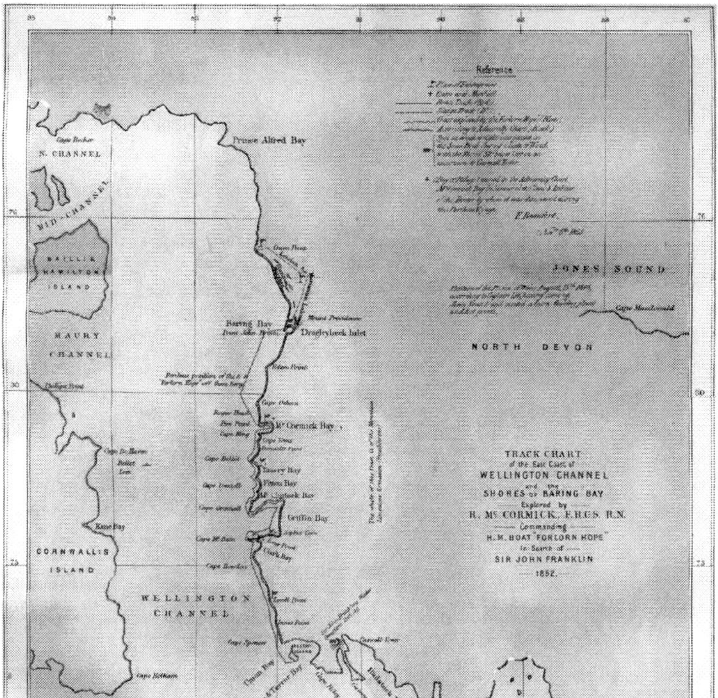

Reference
Prince Alfred Bay
N. CHANNEL
MID-CHANNEL
BAILLIE HAMILTON ISLAND
MAURY CHANNEL
Baring Bay
Dragleybeck Inlet
JONES SOUND
NORTH DEVON
Mc Cormick Bay
Maury Bay
Griffin Bay
CORNWALLIS ISLAND
WELLINGTON CHANNEL
Kane Bay
Maxwell Bay
BARROW STRAIT
TRACK CHART
of the East Coast of
WELLINGTON CHANNEL
and the
SHORES of BARING BAY
Explored by
R. Mc CORMICK, F.R.C.S. R.N.
Commanding
H.M. BOAT "FORLORN HOPE"
in Search of
SIR JOHN FRANKLIN
1852.

Sketched by R. M'Cormick, R.N.

Forlorn Hope running under the lee of two icebergs, aground in the Wellington Channel, in a gale of wind, 23rd August, 1852.

Page 105.

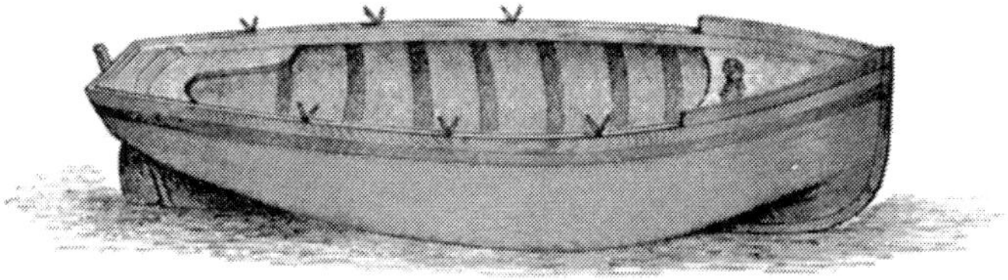

"The Forlorn Hope."

NARRATIVE OF A BOAT EXPEDITION

UP THE

WELLINGTON CHANNEL

IN THE YEAR 1852,

UNDER

THE COMMAND OF R. M'CORMICK, R.N., F.R.C.S.,

IN

H.M.B. "FORLORN HOPE,"

IN

SEARCH OF SIR JOHN FRANKLIN;

WITH

CHART, ILLUSTRATIONS, AND PLANS OF SEARCH.

On Thursday, the 19th of August, 1852, at eleven a.m., I succeeded in embarking upon my long-sought and long-cherished enterprise, in a whale-boat equipped for a month, and manned by half-a-dozen volunteers from her Majesty's ship *North Star*, lying off Beechey Island.

Although it could not be otherwise than a source of the deepest regret to me that the short season for boating operations in these regions was now fast drawing to a close, and with it the more sanguine hopes I had entertained of accomplishing the extended exploration I had contemplated ere the long polar night set in, yet, even in this, the eleventh hour, I was not without a hope of at least setting at rest one question relative to the search, viz. as to the existence of any available communication between Baring Bay and Jones Sound, either by means of an opening or narrow isthmus of land, in the direction of the position laid down in the Admiralty chart, as the spot where a cairn, cooking-place, and footprints, are said to have been visited by a whaler; and have been thought by some, most deeply interested in the fate of our lost countrymen, to have been traces of their wanderings.

This object I fully determined to accomplish, if possible, either by sea or land, even should the formation of "young ice" (so much to be apprehended at this advanced period of the season) form such an impediment as to leave me no other alternative than to abandon my boat, and make my way back to the ship by an overland journey.

At the very moment I was about taking my departure, a sail hove in sight, coming round Cape Riley, which proved to be no less interesting an arrival than Lady Franklin's own little brigantine, the *Prince Albert*, on her return from Batty Bay, in Prince Regent's Inlet, where she had wintered, without finding any traces of the missing expedition. I met her commander, Kennedy, and Monsieur Bellôt, on the floe as they landed, but so anxious was I to make the most of every moment of the brief remnant of the season still remaining, that I had little time to inquire what they had accomplished.

After despatching a few hastily-written lines home by them, I struck across the ice for the floe-edge, where my boat was awaiting me; and hoisting the sail with a strong breeze from the south-west, ran alongside of the *Prince Albert*, standing off and on between Cape Riley and Beechey Island. Hepburn, the faithful follower and companion of the gallant Franklin in his ever-memorable

Sketched by R. M'Cormick, R.N.

Cape Spencer, bearing W.N.W. (all the bearings are magnetic), distant two miles.

Page 106.

journey along the shores of the Polar Sea, was on board this little vessel. I had not seen him since our first meeting in Tasmania, on my arrival there—in the very same *Erebus* of which I am now in search—whilst engaged in the Antarctic Expedition, at the time Sir John Franklin was governor of the colony. In passing so close I could not resist the impulse to jump on board, and congratulate this spirited old veteran with a hearty shake of the hand on his safe return, thus far, from so arduous an undertaking in search of his old commander.

A sudden change in the weather having taken place yesterday, accompanied by a heavy fall of snow, covering all the hills with one uniform mantle of white, too plainly heralding the setting in of winter, rendered my visit a very brief one and shoving off again, we rounded Beechey Island in a snowstorm, and were compelled to lower the sail and pull through some loose stream ice (coming out of the Wellington Channel) to Cape Spencer, where we had our dinner of cold bacon and biscuit, at two p.m.

On doubling Innes Point we fell in with a large quantity of drift ice, setting with the strong current, which runs here from the north-west, rapidly down channel, and apparently extending across to the opposite shore of Cornwallis Land, leaving a narrow passage of open water along the North Devon side, which I availed myself of, pushing onwards between the ice and the land. The shingle beach, between Innes and Lovell Points, is margined by a low glacial formation, giving the latter point a white berg-like termination.

At six p.m. I landed to examine a remarkable conical heap of shingle, not unlike a cairn, as it peered above the snow. It seemed to have been thrown up at the outlet of a watercourse to the sea, the bed of which was now dry; but the cleft in the ridge of rocks through which it passed was roofed over with ice and snow, forming a cavern beneath.

On entering a beautiful grotto disclosed itself, the floor glittering with countless globular masses of frozen drops of water, and the roof with pendant icicles clear as rock crystal. The interior of the cave, which extended to a greater distance than I had leisure to follow it up, was so encrusted over with these aqueous stalactites and stalagmites, that the whole surface sparkled through the faint gleam of light admitted, as brilliantly as if studded with huge diamonds.

The weather suddenly cleared up fine, but the wind shifting round to N.N.W., dead against us, hemmed us in between the ice and the land within a bight, leaving open water in mid-channel, from which we were cut off by a belt of heavy floe-pieces, margined by much sludge, and about half a mile in breadth. Our further progress being thus arrested, we landed at seven p.m. to take our tea, in the hope that by the time that we had finished this refreshing repast a passage might have opened out for us. At eight p.m., however, the ice was jamming us into the curve in the coast closer than ever. I therefore determined on making an attempt to force the boat through it, by poling her along with the oars and boarding-pikes. In this way we succeeded in getting about half-way through, when the swell increased so much as we neared the margin, and the heavy pressure to which the boat was subjected between the larger floe-pieces became so great, that we had to haul her up on the ice, after taking everything out of her, to preserve her from being stove in. We then endeavoured to drag her over the larger pieces, with the intention of embarking the provisions and other things, as soon as she was launched into the loose sludge outside. Whilst thus laboriously employed, the making of the flood-tide augmented the swell and commotion amongst the floe-pieces so much,

Sketched by R. M'Cormick, R.N.

Launching of the boat over the drift ice from Lovell Point Encampment.

Page 108

pressing them together with such violence, that one of the largest and thickest pieces on which we had deposited our provisions suddenly parted in the centre, threatening destruction to everything upon it. It was in a scene like this, and not far from the same spot, that the unfortunate Bellôt was drifted away on a floe-piece of ice, and lost his life, for want of a boat.

In this critical position I was reluctantly compelled to relinquish the attempt for the present, and after landing everything in safety by means of the sledge, we dragged the boat over the floe-pieces and landed her upon the beach. It was midnight before we pitched the tent for the night on a ridge of shingle, after four hours of unceasing, most harassing, and dangerous work, which fairly put to the test the capabilities of every one of my small party, and fully satisfied me that I could not have selected a finer boat's crew for a perilous service, had I had the whole Arctic squadron to have picked them from. After supper, having set a watch for the night as a precaution against a surprise from the bears, whose tracks were rather numerous upon the snow on the beach, the buffalo robes were spread, and all turned into their felt-bags, to enjoy that sound and refreshing sleep, which seldom fails to attend on the wearied and toil-worn, however hard may be the couch or inclement the clime.

Friday, 20th.—The spot on which we encamped last night is a little to the northward of Lovell Point, all around a snowy waste, save and except the narrow shingle ridge on which the tent stood, and that was bare. The northern sky looked black and threatening, not that peculiar dark horizon indicating the presence of open water, and hence technically called a water-sky, but the lurid appearance preceding bad weather; the thermometer during the day rose no higher than 31° Fahrenheit. We saw four large flocks of geese all flying at a considerable height in their usual angular-shaped phalanx, shaping their course for the south, a sure sign of winter's near approach. Saw also many dovekies and kittiwakes and two seals.

On emerging from our felt-bags this morning at six o'clock, in which, chrysalis-like, we had been encased during the night, and quitting the confines of the tent, we found that but little change had taken place in the scene around us: both ice and weather bore much the same aspect. On the outer edge of the ice a heavy surf was still breaking, and large floe-pieces had been stranded on the beach by the heavy pressure in the night. The atmosphere looked gloomy, overcast, and threatening; the thermometer had fallen below 29°, and young ice formed to the thickness of an inch. After our breakfast of cold bacon and biscuit, with chocolate, I took a rough sketch of the encampment, and walked for about a mile along the beach to the northward, in search of a more promising part in the belt of ice for embarkation, but found none, even so practicable for the purpose as the place of our encampment.

On my return, therefore, the boat was once more launched upon the floe-pieces, which, from the wind drawing round more to the westward, had been packed closer together in shore; and at ten a.m., by dint of great exertion, we at last succeeded in gaining the outer margin; but it was noon before everything was got into the boat, having to make three sledge trips from the shore with the provisions and other things. We now launched her into the sludgy surf, where, from her being so deep in the water, although with only a month's provisions on board, and this she could barely stow, her situation was for a few minutes a very critical one, from the risk of being swamped, till by a few lusty strokes of the oars, we were swept fairly out of this vortex of sludge and water into the open channel, and made sail with a

Sketched by R. M Cormick, R.N.

Cape Bowden, from the summit of Cape McBain, W N.W.

Page 110.

fresh breeze for Cape Bowden, going at the rate of about five knots an hour.

In doubling Cape Bowden we had to make a considerable *détour*, to avoid a long stream of ice extending from it to the distance of several miles; and in running through the heavy swell and sludge which skirted it, carried away our rudder, through one of the pintles giving way, which, on examination, was found to have been defective, and the rudder altogether badly fitted. In short, the boat was an old one, which had been knocked about in the late expeditions, and not well adapted for such an enterprise as this. This accident, together with a freshening breeze accompanied by thick weather, snow, and sleet, compelled us to lower the sail at five p.m. I now looked out for a spot to beach the boat, under Cape Bowden, a perpendicular cliff, rising to the height of upwards of 500 feet above the level of the sea; but the extremely narrow strip of shingle beach at its base was so thickly studded with stranded hummocks and berg-pieces of ice, on which a heavy surf was breaking, as to render it alike impracticable either to haul up the boat or find room to pitch the tent afterwards.

On the north side of Cape Bowden we opened a pretty little bay, of semicircular form, most symmetrically so, about a mile in breadth at its entrance, and much about the same in depth; bounded on the north by a low, narrow peninsula, suddenly rising into, and terminating in a tabular-topped cape, about 200 feet in height, separating it from Griffin Bay. We pulled all round the little bay with the intention of encamping there for the night, but found the beach everywhere so hemmed in with a fringe of grounded hummocks of ice lashed by the surf, that not a single opening offered, even for running the boat's bow in between them. A flock of geese, a number of gulls, and several ravens, which we had disturbed in their solitary retreat, took wing on our approach. I gave it the name of Clark Bay; and the headland bounding it to the north, I called Cape M'Bain, after two esteemed friends; the former being one of the few remaining survivors who shared in the glorious battle of Trafalgar, and the latter an old voyager to these regions.

On rounding Cape M'Bain into Griffin Bay, the weather became so thick as nearly to conceal the land, and we had some difficulty in finding a spot where the boat could be beached. After coasting the south side of the bay for nearly a mile within the Cape, we at last succeeded in hauling her up into a little nook between the grounded hummocks with which the whole line of coast was thickly strewed. At 6.15 p.m. we pitched the tent for the night, between two small shingle ridges, lighted a fire, and had tea, with some cold bacon and biscuit.

Griffin Bay presented a most wild-looking scene of desolation; the surrounding hills were all covered with snow; huge masses of old ice which had been stranded by some enormous pressure, lay thickly strewed along its shores, in places piled up in chaotic confusion; and the upper part of the bay was full of loose ice, the winter's floe having very recently broken up. The streams of ice which we met with on our way up channel doubtless came out of this and the adjacent bays.

When about turning into my felt-bag for the night, I found it saturated with water, and preferred taking my rest on the buffalo robe, without any other covering than what the tent afforded, having a black tarpaulin bag containing my change of clothes (all thoroughly drenched by the seas the boat shipped over her bows) for my pillow.

Saturday, 21st.—Rose at five a.m., breakfasted, and started at six o'clock for the summit of Cape M'Bain, on which I found a cairn, containing a small gutta-percha case, enclosing a circular printed in red ink on yellow tinted paper, dated Tuesday, May 13th, 1851, and stating

Sketched by R. M'Cormick, R.N.

Cape McBain, bearing W.S.W.

Page 112.

that a searching party from the *Lady Franklin* and *Sophia*, brigs, had left, for emergencies, on the north point of the bay, a cache of sixty pounds of bread and forty pounds of pemmican. From the spot on which the cairn stands, I took sketches of Capes Bowden and Grinnell, and descended on the south side into Clark Bay, and whilst examining its shores, I saw an Arctic gull and three fine large white hares (*Lepus glacialis*), which, however, were far too shy and wary to allow me to approach within ball range of them: both barrels of gun being loaded with ball, I discharged one after them, which sent them running off at a tremendous rate.

Returning to our encampment, we struck the tent, and after re-embarking everything, made sail with a fair wind from the westward at 9.15 a.m., but still the same overcast and gloomy aspect of the sky. After we had proceeded for some distance, I discovered that a fine musk ox (*Ovibos moschatus*) skull and horns (evidently a bull's from the bases of the horns meeting over the forehead), found by two of the boat's crew, on one of the ridges above the bay, in a ramble they took last night—had been left behind on the beach. This was much to be regretted, as the specimen furnished pretty decisive evidence that these animals must once have existed here, and the probability is that they do so still. It bore evident marks of long exposure to the weather, bleached white, porous, and time-worn.

Standing over for Cape Grinnell, we encountered another heavy stream of ice, which crossed our course as it drifted rapidly out of Griffin Bay, cutting us off from the shore, and we had to get out the oars and pull round it. Our rudder, which we had made an attempt at repairing, again gave way. We passed a shoal of white whales (*Beluga borealis*), and saw the cairn on the point where the depôt of provisions was left. After taking a sketch of the latter, I landed about noon upon a narrow shingle beach, on which we lighted a fire and cooked a warm mess, made of preserved mutton, soup, and potatoes, for our dinner. On walking up the ridge to the cairn, through a heavy fall of snow, we found the provisions gone; and as there were recent footprints up the side of the ridge leading to it, where the melting of the snow had left the soil sufficiently soft and plastic to take impressions, I came to the conclusion that the *Assistance* and *Pioneer* had taken them on their way up channel. Returning to the boat, we shoved off at three p.m., the sun glancing forth a momentary ray through the surrounding murky atmosphere, as we receded from the shore. On rounding the cairn point, we opened another small semicircular bay, strikingly resembling Clark Bay both in size and form; and to which I gave the name of M'Clintock, after my friend, the distinguished Polar traveller, now commander of her Majesty's ship *Intrepid.*

The coast, along which we had now to pull against a fresh northerly breeze, presented a very bold and striking aspect. Bluff headlands, rising precipitously from the water's edge to the height of 600 feet and upwards, and skirted at the base by a narrow belt of shingly beach, profusely studded with stranded hummocks of ice. From the steep fronts of these magnificent cliffs of the mountain limestone projected three or more horizontal tiers of buttresses in strong relief, the effect of which was much heightened by the tiers being bare of snow, and black—so contrasted with their white sides as to give them the appearance of some frowning and impregnable fortress, or imposing battery presented by the broadside of a stately three-decker. Between two of these remarkable headlands, another very symmetrical bay opened out, bounded on the north by a wild, romantic-looking cape, towering upwards with smooth and swelling sides to near its summit, and then abruptly breaking up into angular-

Sketched by R. M'Cormick, R.N.

Encampment at the base of Cape Bellot.

Page 114.

Sketched by R. M'Cormick, R.N.

Cape Daniell, bearing N.W.

Page 112.

shaped rocky fragments, forming a rugged, picturesque-looking crest, seven or eight hundred feet above the level of the sea. To this pretty bay I gave the name of Emery, after an old friend; and to the south headland Cape Daniell, after another esteemed friend; both of whom have their names already enrolled in the annals of African discovery. To the remarkably crested north headland I have given the name of my lamented friend, Bellôt (Cape Bellôt), he having lost his life in its vicinity. I saw the tracks of bears and foxes upon the snow along the beach.

Cornwallis Land, forming the opposite shore of Wellington Channel, piebald with snow, loomed dark and wildly through the mist, at the distance of between twenty and thirty miles, yet I could distinctly make out the point forming its north-eastern extremity. Passed several white whales, a seal or two, and several large flocks of geese, the whole migrating to the south, a few dovekies (*Uria grylle*), fulmar petrel (*Procellaria glacialis*), glaucous and kittiwake gulls.

At six p.m., observing a cairn on a low ridge of shingle, I ran the boat in between the grounded hummocks of ice on the point. Landed and found a tin cylinder containing a notice that the *Assistance* and *Pioneer* had passed on Sunday morning last at ten o'clock, bound to Baillie Hamilton Island—" all well."

From this we had a very prolonged and fatiguing pull along a most dreary line of coast, closely packed with grounded hummocks. The breeze increased to a fresh gale, accompanied by sleet and snow; the thermometer 28°; air cold and pinching, and the whole of the land more deeply covered with snow than any that we had yet passed. The horizon to the north looked black and threatening, and a faint pinkish streak of light seemed to give an additional air of wildness to its aspect. The night, too, was fast closing in, with no prospect before us of the smallest nook where we could haul up the boat in safety till the morning. A long way ahead of us three bold capes appeared in the distance; the nearest, a remarkably black-looking one, prominently jutting out from the snow-clad ridges flanking it on either side. Aground off it was a large mass of ice of fantastic shape, rising from the sea by a narrow neck and then expanding out into the form of an urn, appearing as if filled with white foam rising above the brim in a convex form. A long stream of ice was seen extending out from the Black Cape, which led me to hope that we should find a bay on the other side of it out of which the ice had drifted, and a place of refuge for the night, for my hoat's crew were fairly worn out by pulling for so many hours against a head-sea and strong current (running here, at times, five or six knots an hour), and exposed to such inclement weather.

In passing a low shingle ridge, before we reached the black headland, a cairn upon it caught my eye through the dark gloom in which it was enveloped, and although an ice-girt lee shore upon which a heavy surf was setting, I felt that it was my duty to attempt a landing to examine it. The boat's head was therefore at once directed for the shore, and run in between two heavy grounded masses of ice, leaving just room enough for her bows to enter; the ridge of shingle was too steep to haul her up, or I should gladly have encamped there for the night, unfavourable as was the spot for pitching our tent. We had to walk along the ridge over snow, in some places very deep, before we reached the cairn, and, to our great disappointment, after pulling it down and carefully examining the ground beneath and around it, found no record whatever. It was a small pile of rocks resembling a surveying mark, but when and by whom erected no clue was left upon which to form a conjecture. We saw here recent tracks of bears and foxes on the snow.

Sketched by R. M'Cormick, R N.

Pim Point, S.W. Bay of Refuge. Cape King, S.W. by W.

Page 116.

Sketched by R. M'Cormick, R.N.

M'Cormick Bay (named by the Hydrographer of the Admiralty).

Page 116.

Returning to our boat, after some difficulty in embarking in the swell, the crew, to whom I had given a little brandy each, pulled under its temporary influence with renewed vigour for the Black Cape.

That harbinger of the storm, the stormy petrel or Mother Carey's chicken (*Procellaria pelagica*), the first I have seen during this voyage to the Arctic regions, flew past the boat, and I fired at it, but missed it, the boat rolling at the moment too heavily in the swell for taking anything like an aim. We at last rounded the urn of ice and pulled through the stream, passing between and very close to several huge hard-washed blue masses of ice aground, on which a foaming surf was breaking, and the boat pitched and rolled so much in the ground swell as to ship a good deal of water, compelling us to bale her out.

On rounding the black headland we entered, as I had anticipated, .a fine bay, between three and four miles deep ; but after pulling for some distance along its wild-looking inaccessible southern shore without finding a nook where we could hope to get the boat's head in, being a lee shore, ice-girt, on which a dangerous surf was breaking, we had to pull across to the opposite side, a distance of two miles, the shore of which appeared in the form of low shingle ridges, giving promise of a beach on which we might haul up the boat in safety, as well as a dry ridge, free from snow, for pitching the tent. At first we rowed over a very shallow bottom, upon which the pebbles were distinctly seen, in a heavy ground-swell, but as we neared the north side got into deeper water. It was half an hour past midnight when we at last succeeded in hauling up the boat on the beach between some berg-pieces, which had been forced up by some vast pressure above the ordinary high-water mark.

Whilst some of the crew were employed in getting the things out of the boat, and securing her for the night, and others pitching the tent on the shingle ridge above the beach, which on landing I had selected for the site, the cook for the day lighted the fire, and prepared supper. I strolled with my gun along the ridge round the north point, where huge berg-pieces were piled up one upon another in chaotic confusion to the height of from twenty to thirty feet by some tremendous pressure, occasioned, doubtless, by high spring-tides and heavy north-westerly gales.

The strong breeze we had been pulling against had now increased to a hard gale of wind from the same quarter, accompanied by an overwhelming snow-drift. Thermometer 28°, and piercingly cold,—altogether a dismal night. So that we had encamped none too soon, for our frail boat could not possibly have lived in the sea that was now running outside. Therefore I called the inlet the Bay of Refuge, the black headland I named Cape King, and to the north point I gave the name of Pim, after two enterprising Polar friends, both well known for their enthusiasm in Arctic discovery, and their plans for the rescue of our missing countrymen, in the search for whom Lieutenant Pim, like myself, is embarked in the present expedition.

On my return to the place of our encampment, I "spliced the main brace," that is, served out extra rations, in the present instance, of bacon and Burton ale, to the boat's crew for their supper, after their long day of toil and exposure. It was 2.30 a.m. before we turned into our felt-bags for the night. Mine was, however, still wet, and I lay down on the buffalo rug as on the preceding night.

Sunday, 22nd.—Having retired to rest late last night, or rather early this morning, we did not rise until 10.30 a.m. It was still blowing a hard north-westerly gale, with snow-drift and overcast thick weather; so bitingly cold was the air within the tent, that sleeping, as

Sketched by R. M'Cormick, R.N.

Rogier Head, South.

Page 118.

I always do, at the weather end, where the wind blows in under the canvas, my hands felt quite benumbed throughout the night, from their having been exposed, in the absence of my felt-bag covering. I shaved for the first time since leaving the ship, and made my toilet under the lee of the boat. After our customary breakfast of chocolate, cold bacon, and biscuit, I took from my pocket a little prayer-book, which had been my companion years gone by to both the Poles, North and South, and round the world, from which I read to my boat's crew part of the morning service, finishing with a short, extemporary prayer, which suggested itself at the moment, as best fitting the occasion.

At 12.30 I left the tent, accompanied by three of my men, for the summit of Rogier Head—which I named after an old friend who had been engaged in African discovery—a bold, craggy promontory, about 500 feet in height, overhanging the sea, and about three miles distant from our encampment. Our course lay over some snow-clad ridges up a gradual ascent. At 1.45 p.m. we reached the summit, from which a wide and wild scene of desolation met the gaze; whichever way the eye was directed a grand and sublime spectacle presented itself, to which the fury of the tempest lent an awful interest.

Beneath the precipitous face of the overhanging crag on which I was seated, the surf was furiously lashing the narrow strip of black shingle beach at its base, margined by a belt of shallow water, the limits of which were well-defined by a turbid greenish appearance, contrasting strongly with the dark, very dark, blue colour of the water beyond. Along the edge of this zone of shoal water countless white whales were swimming down channel, literally speaking, in a continuous stream. Amongst them, here and there, one of a piebald colour; and sometimes the back of a straggler or two appearing in the discoloured water itself; all, doubtless, migrating to less rigorous seas, whilst open water afforded them a passage to the southward Overhead a solitary kittiwake (*Larus tridactylus*) hovered with uplifted wing, as it breasted the violent gusts of wind that at intervals swept past, driving along dense volumes of mist from the mysterious north, which came rolling over the dark surface of the channel, on the opposite side of which the bleak and barren snow-streaked cliffs of Cornwallis Land bounded the horizon to the westward, terminating in a black point forming its north-eastern extremity, about the position of Cape De Haven, half concealed in gloom and mist. To the north-westward Baillie Hamilton Island loomed like a dark bank of clouds above the horizon: three or four glistening patches of white light, reflected upon the surface of the dark water through some hidden aperture in the clouds, shone with the brightness of molten silver amid the surrounding lurid atmosphere, rendering the whole scene altogether a fit subject for the pencil of a Claude. We commenced our descent of the mountain at 2.30 p.m., and having taken as careful a survey of the vicinity of our encampment as the thick and unfavourable state of the weather would permit of, reached the tent at 3.30 p.m., without finding any cairn or traces of any one having landed here before us. Only the tracks of a bear and fox on the snow were seen. This morning, on starting, a small piece of drift wood was picked up above the present high-water mark; and last night another musk ox skull was found by some of the boat's crew. It was the skull of a cow, the horns being small, and a space between their bases on the forehead, and not in such a good state of preservation as the last. This gives me sanguine hopes that I may yet fall in with the living animal itself before the winter drives us back to the ship. Discouraging as there is no denying our present prospects certainly are, we must at all hazards solve the Baring Bay problem first. On reaching the tent we found dinner all ready,

Sketched by R. M'Cormick, R.N.

Cape Osborn, bearing N.

Page 120.

and a warm mess of preserved mutton very acceptable. A dismal night; thermometer 25°.

Monday, 23rd.—It blew in heavy squalls all last night. About nine o'clock this morning, however, a lull taking place, I resolved to make an attempt to reach Baring Bay as soon as the sea should sufficiently go down for launching the boat. After breakfast we erected a cairn on the low shingle ridge where the tent stood, and deposited beneath it a tin cylinder containing a record of our visit. The upper extremity of the bay was still covered with the smooth winter's floe, which had not yet broken up.

"MEMO.—A boat expedition from H.M.S. *North Star*, at Erebus and Terror Bay, Beechey Island, in search of Sir John Franklin, encamped here at 12.30 a.m. on Sunday, the 22nd of August, during a gale of wind, and left for Baring Bay on the following morning at 10.30. No traces found.

"R. McCORMICK, Officer Commanding Party.
"*August 23rd*, 1852."

At 10.45 a.m., as we were about to start, I shot a sandpiper (*Tringa maritima*) on the beach. On rounding the outer point we found a considerable swell outside, with a strong breeze to pull against; passing Rogier Head, the headland we ascended yesterday, and another adjoining promontory, we had to contend with a short head-sea, in a deeply-laden boat, with a damaged rudder almost useless, compelling us at times to use a steer-oar in addition to keep her head to the sea, along as dreary and desolate a looking coast as I ever recollect having seen in these regions. The land appeared like a vast wreath of deep snow banked up against the horizon, its continuity broken only by deep gullies in one or two places, with not the smallest bight or indentation along its ice-encumbered shores, on which a heavy surf was breaking, where a boat could find shelter during a gale of wind.

After a most laborious pull of four hours we reached

the steep and almost perpendicular ridge of Cape Osborn, a bold headland of rounded form, white with snow, excepting where a dark blotch appeared just below its summit, formed by the bare rock of the projecting buttresses. This cape may be considered the northernmost boundary of Wellington Strait, which here expands out into the broader Queen's Channel At 1.45 p.m. we passed a very remarkable isolated mass of rock, rising abruptly from the steep face of this ridge about one-third from the summit.

It bore a striking resemblance to the bust of a human figure of burly form, and habited in a cloak and cap; the horizontal layers of limestone rock, of which it is composed, being so arranged as to give the cloak a caped appearance; a slab of the limestone in front of the figure, fancy might liken to a book. This singular specimen of sculpture from the hands of nature, worked out of the rock by the united chisels of time and weather removing the softer portions and leaving the harder standing forth in strong relief—I transferred a facsimile of it to my sketch-book under the name of "Franklin's Beacon," whose attention it could not fail to attract, pointing as it does to those unknown and unexplored regions which lie beyond, around the Northern Pole, untrodden by the foot of man since creation's dawn, and in the deep recesses of which, doubtless, lies hidden his mysterious fate, of which our search, thus far, unhappily has failed to elicit the slightest trace.

At four p.m. we doubled Cape Osborn, on the north side of which a huge pile of dirty yellow-looking old berg-pieces of ice lay aground in the turbid greenish shoal water which skirts the coast all the way to Baring Bay, extending out from the hummock-fringed beach to the distance of a mile or two and upwards, and along which a heavy ground-swell sets upon the shore in a succession of long rollers, through which it would have been utterly hopeless for any boat to have attempted reaching the

Sketched by R. M'Cormick, R.N.

Franklin Beacon, bearing S.E. by E.

Page 122.

land in safety. The coast from Cape Osborn trending round to the N.E. brought the wind more aft, enabling us to make sail, and for some time we made considerable progress, dashing through the heavy cross sea that was running at the rate of five or six knots an hour. Having the breeze with us now, the only chance left us was to run the gauntlet for Baring Bay, in the hope of finding there some haven of shelter after rounding Point Eden, which still appeared at a fearful distance ahead of us; and the long line of foaming crests sweeping over the broad expanse of troubled waters which lie between, threatening to engulf our small frail bark ere we reached it. When we had got about midway between Cape Osborn and this point our situation became a truly perilous one; the boat was taking in water faster than we could bale it out, and she was settling down so much as not to leave a streak free; labouring and rising heavily and sluggishly to each successive sea, so that all expected every moment that she would fill and go down the first sea that struck her, from which only the most careful and watchful attention to the helm preserved her. Fortunately for us, at this critical moment, two small bergs aground providentially appeared on the port-bow, and I immediately ran for them, in the hope of finding the water smooth enough under their lee to enable us, by lowering the sail and lying on our oars, to thoroughly bale out all the water from the boat, which was now nearly full; in this we happily succeeded whilst lying only a few feet from the bergs in comparatively quiet water, protected by their blue hard washed sides from the seas which broke over them to windward, rebounding upwards in foaming columns of serf and spray, which dashed high above their summits from forty to fifty feet in height, presenting a wild scene, at once grand, sublime, and awful.

On again making sail our small over-laden skiff, no longer water-logged, bounded onwards over every sea more freely and buoyantly than before; but as we opened Baring Bay, the great body of water which was setting into it from the broad expanse of the Queen's Channel, with the wind and current both from the N.W., caused such heavy rolling seas to tumble in upon the shore, that our crippled rudder was entirely swept away, and we had great difficulty in steering with an oar a boat so deep in such a sea; and had to get one out on each quarter to keep her head to the sea and prevent her broaching to, when nothing could have saved her from foundering instanter.

The shores all round this bay presented a perfectly flat surface, level with the floe,—which still, though broken up, filled the upper part of it,—and extending to a considerable distance inland, bounded by a slightly undulating ridge of hills in the horizon, averaging, perhaps, 150 feet in height. But one point in these inhospitable shores seemed to offer the faintest hope of a place of shelter. This was a black mount on the south side, of somewhat conical form, having a truncated summit, with shingle ridges in front; and from its marked and conspicuous appearance amid the wide surrounding waste of snow, had particularly attracted my attention from our first opening the bay, as it appeared to me the only spot accessible for ice. It was flanked on the west by an inlet, still covered by the winter's floe. On this spot I had from the first centred all my hopes of finding a harbour of refuge. Putting the boat, therefore, right before the wind, I ran for it through a turbulent ground-swell, over a long extent of several miles of shoal water of a dirty green colour, showing the fragments of rock and pebbles at the bottom on nearing the shore, when two points for beaching the boat offered; one on the port bow, forming a curious natural basin of quadrangular shape, enclosed on all sides by a narrow ledge of black rocks and shingle, excepting in front,

Sketched by R. M'Cormick, R.N.

Mount Providence, Baring Bay, bearing West.

Page 124.

where an opening was left just large enough to admit the boat. Into this, being the nearest, my boat's crew were very anxious to take her, thoroughly worn out as they were by a day of unceasing toil and danger, amid which their cool and manly conduct was beyond all praise. And on losing the rudder and tiller, with which I always myself steered the boat, the ice quartermaster especially proved himself an expert hand at the steer oar at a moment when we were obliged to have one out on each quarter to keep the boat's head to the heavy cross sea that was running, to prevent her from broaching to. I objected, however, to the little land-locked harbour for the boat, on the ground of the chance of being entrapped within it by a change of wind bringing the ice down upon it, and thus preventing us from so readily getting out again; and also from the low, boggy ground, exposed on all sides to the weather, being unfavourable for pitching the tent. I therefore stood on for the Black Mount, ahead, and was fortunate enough to find at its base a sloping beach for hauling up the boat between some grounded hummocks of ice, backed by a shingle ridge, dry and free from snow, on which we pitched the tent at eight p.m., sheltered by another ridge still higher, above which rose the Black Mount. I ascended this eminence, whilst the boat's crew were lighting the fire and getting supper ready. From its summit I saw the ice closely packed all round the bay by the wind now blowing up it, and that this was the only spot where a boat could possibly have found a place of shelter along the whole line of coast, from the bay we left this morning to as far as the eye could reach beyond us to the northward, rendering the navigation of the Wellington Channel extremely dangerous for boats at a late and boisterous season of the year. I saw a flock of geese on the passage here, and another arose from a lake on our arrival. A small fragment of drift wood was picked up on the hill. After spreading all our wet clothes on the shingle to dry, everything in the boat having been drenched with sea water, we had tea and preserved beef for supper, and turned in at midnight, truly thankful to God for our providential escape. Thermometer 27°.

When running under the lee of the berg-pieces aground, at the height of the gale a solitary stormy petrel hovered over the wild scene of spray and foam high above the bergs, recalling to memory the extent of those little cosmopolites' migrations from pole to pole, having met with them as far south as the Great Barrier, suggested to me the following lines:—

TO THE STORMY PETREL.

Pretty wanderer of the ocean,
Ever found in constant motion,
Brief must be thy rest and sleep
On the bosom of the deep.

Skimming in the wake or lee
Of the lone ship on every sea;
Where blows the steady tropic breeze
To where the polar oceans freeze.

I've seen thee on the Antarctic sea,
Beneath the mighty Barrier's lee;
Up Wellington's arctic channel, too,
When searching for lost Franklin's crew.

Without a compass for thy guide
Thou o'er the waste of waters glide,
By some magnetic influence led,
Doubtless, onwards thou hast sped

But restless rover as thou art,
Nature will her laws assert,
And claim from thee a parent's care
Once in each revolving year;

When thou must seek some desert isle,
To lay thy egg and sit awhile,
Until thy callow brood appear
For thee to feed, for thee to rear.

This duty o'er, again thou'rt free
To wander o'er thy much-loved sea,
And teach thy young what course to steer
Under thy parental care.

Far-distant regions to explore,
Known to thy ancestors of yore ;
E'en to the poles thy course may lay,
Following the bright orb of day.

Tuesday, 24*th*.—Breakfasted at 8.30 a.m. As it was still blowing a north-westerly gale, preventing our attempting anything further with the boat, I started at 11.30 a.m., accompanied by one of the boat's crew, on an excursion along shore, with the view of ascertaining the state of the ice, and selecting the best route for sledging round the top of the bay, should a continuance of the present boisterous weather render boating operations wholly impracticable.

On passing a small lake about a quarter of a mile from the encampment, we saw two eider ducks (*Anas mollissima*) with eight young ones swimming on it. I shot the whole of the brood and one of the old ducks ; the other made its escape. Our course at first lay over flat, swampy, boggy ground covered with snow, through which a few straggling tufts of moss, lichens, saxifrages, poppies, and a small species of juncas made their appearance at intervals; the whole intersected by very low narrow ridges of shingle, and a chain of small lakes. The winter's floe had all the appearance of having been recently broken up by the late gales setting a heavy swell into the bay, which had ground it into fragments and hummocks mixed with sludge. A thick fog coming on, accompanied by snow-drift sweeping over the bay from the northward, and concealing the outline of its shores, I struck across the low land for the ridge which bounds it inland, passing several isolated masses of rock, which, as they appeared through the snow at a distance, so much resembled piles of stones artificially heaped up, that, dwelling as our thoughts constantly did on cairns and memorials, we were frequently—until the eye became familiar with these deceptions—induced to diverge from our course to examine them. On ascending the ridge we followed it back to the head of the inlet (south of our encampment), which is nearly two miles deep, and narrow at its entrance, being not more than about a quarter of a mile in breadth, but expanding out to double that width. We walked round several lakes on the ridge of hills, and heard the monotonous, mournful cry of the red-throated divers (*Colymbus septentrionalis*) in the vicinity, but the fog had become so thick as to conceal them from view. On descending from the ridge down a terminal black cliff inland of the tent we had to make head against the gale, which drove the cutting snow-drift in our faces, with the thermometer at 29°. We reached the encampment at five p.m., having only had a shot at a tern, and seen the track of a fox. The ice quarter-master and another of the boat's crew returned soon after us from a ramble round the other side of the inlet, having found the skeleton of a bear.

Wednesday, 25th.—Rose at six a.m.; no improvement in the weather; a quantity of sludge ice driven in shore, which was fast beginning to be cemented together by the formation of young ice, forming an impassable belt for our boat, in front of the encampment. Still too thick and boisterous for boating or sledging. After breakfast I visited the small lake again, and shot three ducks out of a flock of eight young pintails (*Anas caudacuta*). After my return to the tent with them, one of the boat's crew killed the remaining five. We had some of them for dinner, and found them excellent eating. Saw two or three sandpipers, and wounded an Arctic gull (*Lestris parasiticus*), but notwithstanding that the thumb, or tip of the wing, was broken, it succeeded in getting away.

I walked afterwards to the top of the west inlet, accompanied by two of my party, in search of the remains of the skeleton of the bear, they having, on first finding it, brought back with them the skull and pelvis. After a long search we at last hit upon the spot where a rib was projecting from the snow, beneath which we found most of the vertebræ, deeply imbedded in the richest bed of moss we had yet seen, the result, doubtless, of the manure arising from the decomposition of the animal's carcase; although, from the bleached appearance and honeycombed state of the bones, a long series of winter snows would seem to have mantled over them since Bruin dragged his huge, unwieldy frame a few yards above the head of the inlet to breathe his last on *terra firma*, whether in sickness or old age, to become food for the foxes, who had rendered the skeleton incomplete by walking off with most of the ribs and long bones to feast off at their leisure. All that remained I collected, and we returned to the tent through a heavy hail-storm and densely overcast sky, with thick mist, and the thermometer at 25°. Saw some red-throated divers on one of the largest lakes, two tern, and the track of a fox. In the afternoon, the wind shifting round to the westward, and the weather somewhat moderating, though still very squally, I set about making preparations for our sledging journey, the wind now setting directly up the bay, packing the ice so close as to render any attempt with the boat utterly hopeless. Having stowed the sledge with four days' provisions, we dug a trench and made a cache of the remainder of our provisions, filling it up with shingle as a protection against the bears during our absence. The boat was hauled up on the second ridge on which the tent stood, and turned bottom up, with the gear and spare clothes stowed underneath, as a precaution against high tides, which might probably rise higher than usual under the influence of heavy westerly gales.

Thursday, 26th.—I was stirring at three a.m. Morning gloomy and overcast, with snow. Wind round to the eastward and moderated. Thermometer 24°. Walked down to the lake where I shot the ducks; it had frozen over during the night; took a sketch of the encampment from it. Three or four snow-buntings (*Emberiza nivalis*) were flitting about on the ridge above the tent, saluting us with their lively, cheerful note. Yesterday a red-throated diver was shot on one of the lakes by one of our party. At five a.m. I roused out the boat's crew, and we had our chocolate, biscuit, and bacon breakfast.

The progressive fall in the temperature, with the rapid formation of young ice, together with the boisterous north-westerly gales, which had packed the broken-up winter's floe upon the shore in front of our tent, forming a belt of hummocks and sludge half a mile in breadth, and daily increasing in extent, cutting us off from the open water, and requiring only a few calm days to cement it all together, and render the present position of the boat inextricable, were unmistakable signs that the season for boating operations was past; and so soon as a southerly wind from off the land should drive the ice out, no time was to be lost in getting her into the open channel. All, therefore, that now remained to be done was to complete the exploration of this bay by an overland journey.

SLEDGE EXCURSION ROUND BARING BAY.—Having struck the tent, and stowed it on the sledge, with our felt bags, buffalo robes, four days' provisions, and an "Etna" with spirits of wine for fuel, we started at eight a.m.; reached the first low rocky point in the curve of the bay, two miles distant, at nine a.m. Our course lay over the low snow-clad ridges of shingle. From this our encampment hill and boat bore north (magnetic), but here the variation is so great as almost to reverse the

Sketched by R. M'Cormick, R.N.

Encampment, Mount Providence, Baring Bay.

Page 130.

points of the compass. At 9.30 a.m. we struck off more inland, in the direction of the ridge of hills, to avoid a curve of the bay, crossing over a level tract of marshy bog, covered with snow, on which one of the party picked up a small spider. At 10.15 a.m. crossed a rivulet over a pebbly bed, from which some animal was seen on one of the shingle ridges, but at too great a distance to make out whether it was a bear or reindeer, as it disappeared behind the ridge before I could get my telescope to bear upon it. Crossed another running stream, rapidly flowing over its pebbly channel (towards the bay), across which the sledge was carried. I made a considerable détour here in pursuit of the stranger, without seeing anything more of him, and overtook the sledge upon a broad, smooth, snow-clad plain, the monotonous whiteness of which was only broken by the narrow bare ridges and spits of shingle, which intersected its surface like shaded lines, scarcely rising above it. At eleven a.m. my party being somewhat fatigued with this, to them, novel work (and dragging a sledge over the inequalities of land, covered with snow though it be, is a far more laborious task than over floe ice), they had a spell of ten minutes to rest, and take their allowance of rum, mixed with the pure water from an adjacent lake. Saw two sandpipers here, and the track of a reindeer (*Cervus tarandus*), probably that of the animal we lately had a glance of. At 11.30 a.m. reached the head of the curve of the bay we had been steering for; it contained a large patch of loose ice, a low point jutting out from it to the S.E. Point Eden bore N. from this. Passed two small lakes, and heard the cry of the red-throated diver. About noon the breeze died away to nearly a calm, and the men were so heated by their exertions that they took a spell for a few minutes. I saw the land on the opposite side of the Queen's Channel, bearing E.S.E. At one p.m. a portion of the spine of some animal was picked up; saw two more sandpipers, and passed another lake. The breeze springing up again, in less than an hour had freshened to a gale, accompanied by a sharp snow-drift, which swept like volumes of smoke over the wide waste around us to the sea, which was scattered over with streams of hummocky ice. We rested for an hour to dine, on the side of a low shingle-ridge, having the bay in front, a lake on either side, and another in the rear, from which we drank delicious water, with our cold bacon and biscuit meal. Started again at three p.m.; I shot a tern (*Sterna arctica*) near a small gap or pass, in an embankment here, skirting the bay. At 4.15 p.m. Point Eden bore N.N.W., and a peak of the land on the opposite side of Wellington Channel, S.E.

At 5.30 p.m. filled our kettle with water from a neighbouring lake, and having boiled it over the spirit lamp of the "Etna," made tea under the lee of the sledge, in the midst of this wilderness of snow. Cape Osborn with Eden Point bore N.N.W. At 6.30 p.m. started again, and at seven p.m. when some distance ahead of the sledge, pioneering the way, as was my custom, I came suddenly upon the track of the musk ox, close to one of those numerous running streams, by which the chain of lakelets studding these marshy flats empty themselves into the bay. The animal appears to have attempted crossing over the frozen surface of the stream, but finding that the ice, which was broken by his two fore-feet, would not bear his weight, retreated, crossing his own track in the direction of the hills bounding the horizon to the southward. From the appearance of the footprints (which measured five inches, both in length and in breadth) it must have passed very recently, as there was a driving snow-drift at the time, which would soon have effaced the impressions. These footprints, when taken in connection with the two skulls recently found, afford, I think, indisputable evidence that the musk ox is an inhabitant of North Devon, at least during the summer months, and is, probably, now migrating to the southward for the winter. But their course thitherward, and how they get across Barrow Strait, is not so easily explained; they must, at all events, wait till the strait is frozen over.

The black point, with its rounded snowy top, in which the ridge of hills environing the bay terminates to the northward, and which we have had in sight so many hours as the goal to be reached before we pitched the tent for the night, has for several miles appeared at the same distance, or, as the sledge's crew would have it, receding, as mile after mile, with weary and jaded steps, they toiled along, dragging after them the cumbrous sledge, and still the dark point appeared no nearer. Fairly exhausted, they were compelled to take more frequent spells to rest for a few minutes. The night, however, looked so threatening, the northern sky intensely black and lowering—premonitory signs of the wind going back to its old stormy quarter—that I was very anxious to secure the shelter of the point ahead for pitching the tent under; as in the exposed, wide, and bleak waste around us, the canvas and poles supporting it would scarcely have withstood the violence of the strong gusts of wind.

The dark sky was preceded by a very remarkably-tinted horizon in the north, in which streaks of a fine olive-green, alternating with bands of an amber colour, and a rich chestnut-brown zone, intersected horizontally; the side of the hills about Prince Alfred Bay, crested by a dark neutral tint, vanishing into a leek-green. When within about a mile of the point, to encourage my sledge-crew and convince them that we were in reality now drawing near it, I walked on ahead at a quickened pace and ascended to the summit; and, on descending again to the extreme rugged point, I found them pitching the tent on the shingle-ridge beneath. It was exactly midnight, and thick weather with fine snow. A fire was soon lighted, tea prepared, and bacon and biscuit served out for supper. It was nearly two o'clock in the morning before we turned in, all thoroughly knocked up with the day's exertions.

Friday, 27th.—Morning overcast; I left the tent at eight a.m., and whilst breakfast was preparing, ascended the rugged point above our encampment to get a view of our position. At first scrambled over a confused pile of rocky fragments, swelling out above into a broad, smooth, and round-backed hill about 300 feet in height, commanding a view of the shores of the curve of the coast to the northward of it, laid down in the chart as Prince Alfred Bay; an isolated peak, apparently some little distance inland, just showing itself over the highest range of hills on the north side; this ridge terminating in two black table-topped, bluff headlands, running far out to the westward; but the horizon was too hazy for making out distant objects sufficiently clear for getting the different bearings correctly, which, as this spot promised to be the extreme limit of our journey, I was the more anxious to obtain before I commenced my return, more especially as the sun had been hid from our view by fogs, mists, and constantly overcast skies, accompanying the tempestuous weather which has attended us in all our movements since we left the ship; so that no opportunity has offered for getting observations for the latitude and longitude, and consequently my little pocket sextant has remained idle in its case.

In the hope that the weather might clear up about noon, I returned to the tent to breakfast, having seen only about half a dozen snow-buntings flitting about the hill-top. My party were glad to take a siesta in the ten

Sketched by R M'Cormick, R.N.

Westernmost Bluffs of Alfred Bay, bearing E. by S., and Peak, E.S.E., as seen from the summit of Owen Point.

Page 134

Sketched by R. McCormick, R.N.

Owen Point, distant five miles, bearing E.S.E.

Page 132.

to-day, so knocked up were they after their laborious and toilsome forced march of yesterday, dragging a heavily-laden sledge over a distance of about thirty miles, having actually travelled this within the space of sixteen hours at the average rate of rather more than two miles in an hour, resting for dinner and tea an hour at each meal; the longest sledging journey by far, I believe, that has yet been accomplished in one day without the aid of dogs.

At 1.30 p.m., during a temporary clearing away of the mist, I again ascended the hill above our tent, bounding the low shores of Baring Bay on the north, which I have named Owen Point, in honour of my friend Professor Owen, the distinguished naturalist and Cuvier of our own country, who has evinced a lively interest in the Franklin search and Polar discovery.

Baring Bay, indeed, scarcely deserves the name of a bay; it is little more than a broad sweep in of the coast, and is so shoal on entering it from the southward, that I could see the pebbles at the bottom for several miles off shore; and had good reason to remember the heavy ground swell that rolled over it in surges, threatening destruction to the boat every minute, in the gale which drove us before it, to seek the only place of shelter which the whole length and breadth of its shores afforded under the Black Mount.

A black table-topped bluff, bearing E. by S. by compass, forms the western-most extremity of Alfred Bay, on the north side; and a little to the eastward of this, peering just above the high ridge of land, is a peak bearing E.S.E., and being the only apparent peak, would therefore seem to be Mount Franklin, as there is no hill whatever representing it in the position in which it is laid down in the chart inland of Baring Bay. A line of hummocks of ice as if aground, appears in Baring Bay, about two leagues from shore, which may possibly cover a shoal or very low islet. Distant land in the Queen's Channel, apparently Baillie Hamilton Island, &c., bore from N. by E. to N.E. by E. Cape Osborn bore N.N.W., and the Black Mount above our boat N.W. by N. Whilst taking a sketch of the bays and distant points, the ice quartermaster and some of the boat's crew meantime erected a cairn on the north side of the hill, the others being employed cooking dinner, &c., preparatory to our departure. We finished the cairn at 3.30 p.m., and placed beneath it a tin cylinder containing a record of our proceedings thus far. On descending the hill we discovered an ancient Esquimaux encampment on its acclivity, consisting of a pile of fissile rocks of semicircular form in front of a natural wall of the stratified rock which jutted out from the side of the hill. We dug beneath it, but found nothing. The rock, a dark brown-coloured limestone, highly crystalline, and the surface embossed with the elegant scarlet lichen (*Lecanora elegans*). On our return we had our usual meal of cold bacon and biscuit, with some tea.

At five p.m. we struck the tent to commence our return to the boat, the state of the weather unfortunately precluding any astronomical observations being taken for fixing the positions of the land, which have evidently been laid down much in error in the chart. Passing one of the largest lakes I had several shots at a pair of red-throated divers; they had a young one on the lake, which I shot, and started again at 6.20 p.m. We encamped for the night in the midst of the unsheltered waste of snow, nearly half-way back to our boat, at 10.30 p.m.

Saturday, 28th.—At eight a.m. breakfasted, struck the tent, and started again at 9.30 a.m. This was about the most uncomfortable night we had yet passed, blowing a hard gale of wind, accompanied by a fall of snow, and clouds of drift, and so cold that we could not get warm

Sketched by R. M'Cormick, R.N.

Mount Providence, with Eden Point and Cape Osborn, W.N.W., as seen from the summit of Owen Point.

Page 136.

all night. The wind finding its way under the tent, shaking it so violently that we expected every moment the poles would give way, and the canvas come down upon us for a coverlet. The thermometer stood at 29°. The watch during the night heard a distant sound, like the bellowing of cattle. Probably the musk ox whose footprints I fell in with yesterday, but concealed from view by the ridge of hills inland of us; for sounds may be heard at a great distance, in the highly rarified state of the air in the still solitudes of these regions. This snowy desert was here and there dotted over with boulders of rock, richly ornamented with the beautiful and bright scarlet lichen, and intersected by numerous rivulets and lakelets, some of the largest of which were now half frozen over; and the ice on the less rapid fresh water-courses permitted the sledge being quickly drawn over by the whole party without breaking. At eleven a.m. we rounded a deep curvature in the shores of the bay, the wind edging round to its old quarter in the N.W., snowing with a strong drift. Saw three or four tern, whose vociferous clamour over our heads proclaimed their anxiety for the safety of their young, evidently not far off. At intervals we heard the wild deep-toned and mournful cry of the red-throated diver rising from some adjacent lake, music to the ears of us lone wanderers, in the dearth of life and sound around us. We saw one large flock of ducks only, going south. At 11.30 a.m. crossed an elbow of the low shores, forming a considerable convexity in the bay, from which a deep curve ran up beyond it; bounded on the west by a low black point, covered with broken-up fragments of limestone, faced with the scarlet lichen, and abundantly fossiliferous, more especially in corallines, of which I collected some specimens. Here we became enveloped in a thick fog, which, with snow, continued till we reached our old encampment. At two p.m. crossed a patch of loose dark sand,

and the sledge-party rested for a few minutes near a rapid stream, after crossing which the sledge soon came upon its outward track of yesterday. Saw three or four sandpipers, and wounded an Arctic gull; which, falling somewhere in a dark shingle watercourse, about a quarter of a mile from where I shot it, I lost, after making a considerable *détour* from the sledge's course in search of it, for I have not yet been able to obtain a specimen of this solitary bird, mostly met with singly or in pairs, and of which we have seen only three or four individuals throughout our journey, all very shy and wary. On coming up with the sledge, we were drawing near the Black Mount, and I proceeded on ahead of my party to see if all was right. Reached the boat and *cache* at four p.m. in the midst of a snow-storm, with the wind at N.W. Found everything as we left them, with the exception of the gratifying sight of open water in the cove; all the ice having been driven out during our absence, by the southerly winds, which blew for a few hours, leaving only a narrow belt of loose sludge near the beach, and no impediment in the way of getting to sea in the boat. It was just low water, and the large urn-shaped masses of ice were left high and dry in hollows in the bed of shingle which they had made for themselves in the ebb and flow of the tides, and to the repeated action of which they owe their hour-glass form. On the arrival of the sledge we pitched the tent on the old spot. A large flock of ducks alighted in the bay this evening.

Sunday, 29th.—We did not rise until eight a.m. This is the finest morning that we have experienced since leaving the ship; and all our clothing and bedding being so saturated with moisture as to prevent any of us from sleeping last night, I took advantage of the favourable change in the weather to have everything spread outside the tent to dry. Being Sunday, I determined to make it a day of rest to recruit the exhausted energies of my

Sketched by R. M'Cormick, R.N.

Owen Point Encampment, North Point of Baring Bay, bearing E.S.E.

Page 134.

men before we commenced our homeward voyage; all still feeling more or less the effects of the fatigue attending their unremitting exertions for the last two days; one evincing a slight disposition to snow blindness, and another some dental irritation.

After they had all had the great comfort of an ablution and shave, I read part of the morning service to them in the tent. Our dinner, as yesterday, consisted of a warm mess of preserved mutton, soup, and potatoes, with Burton ale. Wind round to the westward, breaking up the winter's floe in the inlet west of the encampment, and which was rapidly drifting out past us. The rise and fall of the tide here is considerable, some six feet, probably. The wind this evening shifted to the N.W., with a fall of snow in large flakes. Night overcast and misty, with a black-looking horizon to the northward. We turned in at nine p.m.

Monday, 30*th*.—I was up this morning and outside the tent as early as four o'clock to look around, and having well weighed both our present position and future prospects, to determine on the best course to be adopted; when, taking into consideration the advanced period of the season and unpromising appearance of the weather, that nothing further could be accomplished in the search northward and eastward of this bay, I very reluctantly decided on returning to the ship, and we commenced stowing the boat and making preparations for our return.

At 9.30 a.m. we erected a cairn on the summit of the Black Mount, which I called Mount Providence, in commemoration of our providential deliverance from as perilous a position as a boat could possibly have escaped from—placing beneath the cairn a tin cylinder, enclosing a record of our proceedings, of which the following is a copy:—

"MEMORANDUM.—A boat expedition from her Majesty's ship *North Star*, at Erebus and Terror Bay, Beechey Island, in search of Sir John Franklin, arrived here on

Monday, August 23rd, at midnight, during a gale of wind and heavy sea, which carried away the rudder of the boat and nearly swamped her.

"On Thursday last, sledged on the snow over the low lands round the head of the bay, without finding any opening to the eastward or traces of the missing expedition, returning to the boat on Saturday afternoon. Weather during the preceding week has been most unfavourable, blowing, snowing, and foggy, with the thermometer constantly below the freezing point. The lakes frozen over, and every appearance of winter rapidly setting in.

"Launched the boat this morning on the making of the tide, to return down Wellington Strait and examine the bays along its eastern shores. A memorandum of our sledge-journey has been deposited under a cairn erected on the summit of the northern point of the bay.

"R. McCormick, Officer Commanding Party.

"*Monday, August 30th*, 1852."

To the inlet running up on the west side of Mount Providence, from S.S.W. to N.N.E., I gave the name of Dragleybeck, in commemoration of the birthplace of Sir John Barrow, Bart., and in compliment to his son, John Barrow, Esq., of the Admiralty, F.R.S., who, following up his father's career, has earned for himself a distinguished position in the history of Arctic discovery by his noble and unceasing efforts in furthering the search for the brave but ill-fated Franklin and the rest of our long-lost countrymen. And with the point terminating the inlet I have associated the name of a dear old friend, now no more, John Brown, Esq., F.R.G.S., the distinguished author of the "North-West Passage," and the plans for the search of Sir John Franklin.

The chain of lakelets I named, after my two sisters, the Louisa and Marianne Lakes, and the level tract of land in which they are situated, the Runham Marshes.

Sketched by R. M'Cormick, R.N.

Franklin's Beacon, S.S.E.

Page 140.

Descending to the ridge, which is about fifty feet above the beach, and from thence to the lower one on which the tent stands, we struck it and erected another cairn on the spot where it stood. The rocks here are sparingly fossiliferous. It was a very low tide this morning, being out 100 feet from the last high-water mark.

After a luncheon of cold bacon and ale, to fortify the boat's crew for the long pull they had before them to the next bay, against a head-wind and pinchingly cold air, we about noon launched the boat between the heavy hummocks of ice aground, five or six feet in height when high and dry. Had snow, fog, and mist, with a short head-sea to buffet with; the drops of water froze on the blades of the oars as they rose from the sea after each stroke, and accumulating till the lower edges became fringed with pendant icicles; the water shipped over the bows soon froze at the bottom of the boat, so that had there been much sea on we should soon have had a very dangerous kind of immovable, glacier-like ballast.

We saw a seal or two, a flock of ducks, a few dovekies, fulmar petrel, and the Arctic gull. At 6.45 p.m., on rounding Eden Point, the trending of the coast in a S.W. direction enabled us to make sail. We carried away a temporary rudder which we had constructed just before we started out of the head of a cask from the wreck of the *M'Lellan*, American whaler, lost by the unfortunate but enterprising seaman, Captain Quayle.

At 7.30 p.m. we doubled Cape Osborn, and, a quarter of an hour afterwards, Franklin's Beacon, standing forth through the mist in strong relief from the side of the ridge. At 9.30 p.m. reached our old place of encampment in Refuge Bay. Found much more snow here than when we left it last, being very deep in places. Pitched the tent close to our cairn, snowing all the time, and pinching work to the men's fingers. Thermometer 27°. The state of the tide prevented us from hauling up the boat on the shingle ridge, which for greater security I always get done if possible; we were therefore obliged to let her ride in the cove, with an anchor out on shore.

Tuesday, 31st.—The morning's dawn brought with it the same kind of weather as yesterday—snow, mist, and fog. Rose at 6.30 a.m. The first fox (*Canis lagopus*) was seen by the watch last night near the boat; represented to have been of a brown and white colour. I found my aneroid barometer this morning quite useless, having sustained some injury from being thrown on the beach in the clothes-bag in clearing the boat, and into which it had been accidentally put. This was a serious loss to me, as I had intended measuring all the heights with it in returning down channel.

From the point I took sketches of the two capes south of the bay, together with the opposite coast of Cornwallis Land.

At 2.45 p.m. having embarked everything, we pulled all round the bay, closely examining its shores, and landing at all remarkable points. At about half a mile from the top got soundings in thirteen fathoms, and within less than a cable's length of the shore the soundings gave four fathoms very regularly. The winter's floe had not yet broken up in a creek at its south-western extremity, and young ice had formed here to the thickness of four inches. This is the only safe and well-sheltered bay along the whole of this coast for anchoring a ship.

Saw several seals, gulls, and dovekies, and shot one of the latter. Landed at a little cove for a few minutes to examine the rocks, and sounded again, still getting four fathoms. At 5.30 p.m. landed near a black cliff in a thick snow-storm, and examined a remarkable-looking ravine running up from it.

6.30 p.m.—Had to pull through a quantity of sludge

Sketched by R. M'Cormick, R.N.

Coast of Cornwallis Land, on the opposite side of Wellington Channel, as seen from the Bay of Refuge.

Page 142.

Sketched by R. M'Cormick, R.N.

Cape Toms, bearing S.W.

Page 146.

ice round the outer point in clearing the bay. Took a sketch of the headlands and entrance to the bay from the southward. Shot at and struck a seal, but he escaped us. Saw four or five ducks.

At 8.30 p.m. doubled the next cape, to which I gave the name of Toms, after my friend the assistant-surgeon of the *North Star*, an enterprising young officer. At 9.30 p.m. passed the point where the cylinder and memoranda were found coming up channel, which I called Domville Point, after my friend and brother-officer, the surgeon of the *Resolute.* About 10.30 p.m. entered Emery Bay, and encamped on a fine hard shingle ridge.

Wednesday, September 1st.—I was awoke between three and four o'clock this morning by the ice quarter-master—who had the watch—running into the tent, and reporting that our boat was swamping in the surf by a sudden squall coming on with the flood tide. On hastening down to the beach I found her broadside on, and half full of water and sand. On getting her round, head to sea and stern in shore, we succeeded, after some labour and a thorough drenching, in getting everything out of her, and hauling her up above the shingle ridge, capsizing her as soon as she was sufficiently clear of the breakers, to empty out the water and sand. It now blew a hard north-westerly gale; the sky was densely overcast, and the air pinchingly cold: thermometer 29°.

Breakfasted at eight a.m. The boisterous state of the weather not affording the slightest prospect of our being able to make a move to-day, with such a sea running outside; therefore, after drying the things, and repairing the damages sustained by the morning's disaster—fortunately nothing of a more serious nature to our provisions than the soaking of an ullage of biscuit in salt water—I planned an excursion round the inland ridges of hills; and to spread the search more widely, separated our party into three divisions of two men in each, taking our guns in the hope of meeting with some game, such at least as these desolate shores have to offer. We started at five p.m., leaving only the cook for the day in charge of the encampment, and a gun to defend himself from bruin, should it be needed. I directed one division to ascend the ridges south of the bay, another directly inland, whilst myself, accompanied by one of the boat's crew, proceeded up the hill to the north; first, passing through a romantic-looking, deep, and narrow ravine, with steep precipices on either side formed of limestone rock, banded horizontally in places with veins of gypsum three or four inches in thickness. I entered this ravine last night, whilst supper was getting ready, and traced the foot-prints of a fox to his domicile in the rocks; but saw nothing of him this morning. Following a zigzag course for about a mile, the black crags breaking through the white mantle of snow which now deeply covers the land, gave it a very picturesque appearance, terminating in an open space between the hills. On emerging we ascended the hill bounding it on the right, and shaped a south-easterly course, so as to fall into the track of the other divisions of our party on the central ridge. On sighting them we descended to the shores of the bay, examining the beach all round to the encampment, without meeting with the slightest traces or indication of any one having preceded us here, and not a living thing to break the death-like stillness and utter desolation of the scene. We reached the tent at seven p.m., and the other parties returned soon afterwards with the same results.

When on the highest ridges I carefully observed the appearances of the land in an easterly direction within the extent of vision, for any apparent break of continuity that might afford an indication of water beyond, never losing sight of the possibility of Jones's Sound sweeping

Sketched by R. M'Cormick, R.N.

Domville Point, N.

Page 144.

round in its course near the heads of some of the deeper inlets of the Wellington Channel, taking a westerly course from Baffin's Bay in the direction of Baring Bay, as Jones's Sound is represented to do in the chart. But neither Baring Bay nor the two other bays that I have since closely examined afford any indication of the vicinity of open water. An intermediate series of ridges of hills, one just rising above the other, and for the most part running parallel with the coast, bound the tops of all the bays; and I have never seen the gulls or other sea-fowl flying inland to the eastward, although I have at all times watched them narrowly in their flight.

Thursday, 2nd.—Breakfasted at seven a.m. The violent gusts of wind, accompanied by heavy snow-drift, during the night, nearly blew down our tent, and the air was excessively cold. Anxiously waiting for the gale to abate, to proceed on our voyage. After sketching the encampment and the adjacent hills, I walked up the ravine, and filled a haversack with specimens from the gypsum vein. Dined at 1.30 p.m., and built a cairn near the tent, under which I deposited a tin cylinder, enclosing the usual record of our proceedings. The gale abating during the day, as soon as the sea had sufficiently subsided I took advantage of the temporary lull to start at four p.m., notwithstanding a dark threatening horizon, with the hope of reaching our old quarters in Griffin Bay before we should encounter a second edition of the gale, which it was but too evident was brewing up in the north. On starting, saw a solitary snow-bunting on the beach. We sounded in crossing the bay with a line of twenty-three fathoms, and no bottom at this depth. The ridges round the top of the bay have a mean height of about 200 feet.

I sailed round the next semicircular bay, which I called Fitton Bay, after an old friend and distinguished geologist, Dr. Fitton (who was the first to direct my attention to the structure of that highly interesting and remarkable island, Kerguelen's Land—Desolation Isle of Captain Cook—in the southern seas). Closely and carefully examined its shores and ridges, and got soundings in from four to five fathoms at 100 yards from the beach. There is no shoal water in either of these bays, both being deep. The boat got into heavy rollers outside of the headlands; one or two of which struck her, filling us with more water than we needed, having had enough of that element already. A black threatening squall rising to windward, we exerted every effort to reach Griffin Bay before it overtook us. At 5.30 p.m. we rounded Cape Grinnell in a snow-storm, into smooth water. Saw the provision cairn on the point, and two seals swimming. Sailed close in shore round the bay, which is margined by shingly beach with hummocks of ice aground all round (as usual on all these shores), backed by a ridge of hills from 100 to 200 feet in height, receding inland in the form of an amphitheatre.

On first rounding the north point, an arm of the bay runs into the N.E.; here we passed a snug little creek enclosed in the shingle banks, leaving an opening just sufficient for admitting a boat, secure from ice and weather; but having a fair wind, I was anxious to make the most of it, inauspicious as was the aspect of the heavens.

We reached the top of the bay, which is about six miles in depth, at seven p.m., and found a low shingle and mud flat, backed by boggy ground, and extending inland to the base of the amphitheatre of hills, interspersed near the beach by pools of water, which appeared to be full of small fish, as the gulls were far more numerous here than at any other spot we have yet visited. A large group of kittiwakes and fulmar petrel, with an ivory gull or two amongst them, were evidently making a good harvest, repeatedly rising with a fish about the size of a pilchard in their beaks after each rapid downward plunge in the water. A solitary Arctic gull was actively carrying on at the same time his buccaneering depredations amongst them whenever an opportunity offered for robbing an unlucky gull of its prey, by compelling it to drop the fish with a scream, which, with great tact, was caught by this sea rover before it dropped into the water.

I ran the boat's head in, but the water was so shoal that she grounded at too great a distance from the beach to effect a landing; and just as I was about stepping out at a more favourable spot, a little further on, with the intention of shooting some of the birds and obtaining specimens of the fish they had swallowed, a bear was discovered on the floe which filled up the inlet at the S.W. corner of the bay. Bruin being considered by all hands, and certainly not the least so by myself, higher game than the gulls, the sail was hoisted instanter, and the boat's head in a few minutes was dashing through the swell (which was now setting into the more exposed part of the bay) before the wind in the direction bruin was leisurely pacing along the ice, on the look-out for a seal, several of which were swimming about the bay. Before we reached the floe, which was of young ice already six inches in thickness, he had, however, taken alarm, and made off for the land, disappearing behind a point jutting out from the inlet.

Finding that the squall which had been threatening for some time was now coming in good earnest upon us, I brought the boat's head round for the south headland of the bay, the site of our former encampment upon the way up channel, in a sheltered cove a little within the headland; but as we became more and more exposed to the sea setting into the bay, in a boat so deep in the water, and so leaky from one of her planks having been stove in by the ice in the bad weather we had been incessantly exposed to, the water from the leak, together with the occasional shipping of a sea, so gained upon us, notwithstanding that a hand was kept unceasingly baling her out, and having no rudder, we had to bear up for the nearest land to us, distant nearly two miles, although unfortunately a lee shore, on which a heavy surf was breaking. We got soundings in twelve fathoms, and saw a second bear. Selecting the most favourable spot that offered for beaching the boat, in a curve of the coast somewhat protected from the rollers by a low point, we backed her in stern foremost, letting go the anchor over the bows, and running a line out astern to the shore so as to keep her head to the sea till everything was got out of her, and fortunately landed without sustaining any damage from the surf, which was breaking heavily against the boat's quarter, save and except a drenching to ourselves.

Before we had hauled her up between the masses of ice into a place of security for the night on the shingle beach, the thermometer fell as low as 25°. The air was bitingly cold, and snowing all the time.

After pitching the tent on a fine hard shingle ridge, clear of snow, the fire lighted, and supper, with a cup of warm tea, under cover of the canvas, we turned into our felt-bags for the night, and soon forgot our toils in a sound sleep.

Friday, 3rd.—Passed the most comfortable night that we have yet had, the ground being hard dry shingle on which our buffalo robes were spread. We were confined within the tent all day by stress of weather, which has been most winterly; blowing, as usual, a hard north-westerly gale, with heavy snow-drift, half burying the tent, the sky overcast with a dense mist, and continuous fall of fine snow. Thermometer throughout the day standing as low as 26°, and the air piercingly cold. The fire outside of the tent took double the usual time in

Sketched by R. M'Cormick, R.N.

Encampment in Griffin Bay.

Page 148.

boiling the kettle; and the pemmican which we had for dinner to-day, for the first time since we left the ship, was hard frozen when taken out of the case.

I had a shot at an eider duck which alighted in the bay. A few glaucous gulls (*Larus glaucus*) flew past the little inlet, which I named Sophia Cove. I occupied myself this evening with my plans of search. Had the last of our Burton ale to-day, and turned into our sleeping-bags at about ten p.m.

Saturday, 4th.—Weather much the same as yesterday, prevented us from putting to sea; but, as the wind and snow-drift had somewhat abated, I formed a party for a bear-hunting excursion to the top of the bay, when just as we were getting our guns ready for starting, bruin himself anticipated our purpose by suddenly making his appearance, and thus saved us a day's buffeting with this inclement weather. One of the boat's crew having reported him in sight, on going outside of the tent I saw a fine full-grown bear (*Ursus maritimus*) sauntering leisurely along the beach, about midway between us and a point towards the entrance to the bay, to which I gave the name of Bear Point. As his course was direct for the encampment, I ordered my party within the tent, to avoid alarming him, whilst I watched his movements from the door. Bruin, however, evidently suspecting that all was not right, suddenly altered his course to pass inland of the tent, at the back of the shingle ridge above it. The instant he disappeared behind the ridge, I made direct for it, to intercept him, desiring my party to be ready with their rifles to cut off his retreat should he happen to escape the fire from my old double-barrel, which had, a quarter of a century before, been fatal to bruin's race in the island of Spitzbergen. On my rising the ridge, bruin turned his head inland, when, after firing both barrels, the ball from the second one brought him on his haunches, at the distance of sixty yards from me.

It was only for an instant, however, for he gathered himself up again, and retreated towards the beach, evidently mortally wounded; and after running the gauntlet of a whole volley of balls from the rifles and muskets of the boat's crew, who, being too eager and excited, I suppose, fired so hurriedly that not a ball took effect; and under their fire he took to the water, swimming out into the bay for the distance of two or three hundred yards, when he wore round with his head in shore, unable any longer to make head against the wind, which was blowing dead on shore. His last efforts to struggle against it must have been desperate, for he had no sooner borne up than his huge form floated on the water a lifeless mass, just as I was about launching the boat to go in pursuit of him. After a short interval the wind drifted him on shore about 200 yards from our encampment, to which we bore him on the sledge; and, cold as it was, we set about skinning him. immediately; when, strange enough, we found on examination that my second ball was the only one that had struck him, entering about a foot above the insertion of the tail, and an inch on the left side of the spine, literally drilling him through, and making its exit by the mouth, splintering two of the canine teeth as it passed out. As a proof of the extreme tenacity of life in these hardy creatures, this animal had one of the largest internal arteries divided by the ball in its course, which poured out so much blood that it was streaming from his mouth and nostrils in such a torrent as to dye the surf around him of a deep crimson colour as we hauled him up on the beach, and on opening the body a deluge of the crimson fluid flowed out. Yet with this deadly wound he managed to run at his usual speed about 200 yards to the beach, and then swim against a head sea for at the least as great a distance farther, making fearful struggles until the moment of his last gasp for breath.

He measured seven and a half feet in length, was finely moulded, and in excellent condition. We had a rump steak off him, as an addition to our pemmican dinner, and found it infinitely better eating than some of the beef I have tasted which had been supplied the ship. At midnight the wind veered round more to the north, with a dark horizon in that quarter. Thermometer 26° Fahr.

Sunday, 5th.—No change in the weather, boisterous as ever, and thermometer at 26°. Had bear steaks for breakfast. Read part of the Morning Service to my party in the tent. Saw several seals swimming about the bay, and another bear on the floe at its upper end, but not within our reach: I could just make him out with the aid of my telescope. An ivory gull (*Larus eburneus*), showing great confidence, hovered about the remains of bruin during the greater part of the day, apparently enjoying a most sumptuous feast. Several glaucous gulls shyly hovered over in passing by, but did not venture to alight; saw also a solitary snow-bunting.

Night threatening, with a black and lurid sky, still blowing hard, with much surf in the bay. Wind shifted round to its old quarter in the N.W. again, with the thermometer down to 24°, and bitterly cold.

Monday, 6th.—Rose at six a.m. Wind more off the land and somewhat moderated, with less sea on outside; the young ice at the upper inlet of the bay which had been broken up by the swell setting on it during the gale, was drifting out past us in considerable quantity, forming a belt along shore.

Commenced preparations for shifting our encampment into the next bay, as soon as the swell along shore subsided sufficiently to enable us to get the boat afloat, and round the headland, the vicinity of which, and summit of Cape Bowden, I was anxiously desirous of more thoroughly examining than my time permitted of when outward bound. Erected a cairn upon the ridge where we had encamped, and deposited beneath it a cylinder containing a record of our proceedings.

At 10.30 a.m., on the wind and sea going down, we launched the boat, and had to row through sludge and brash, intermixed with hard floe pieces of the bay or young ice, which so impeded the progress of the boat that the crew had a most laborious hour's pull in getting through little more than a mile of it.

We landed at our old place of encampment *en passant* to look for the musk-ox skull which we had accidentally left there. But the change which the place had undergone during our absence had been so great that we could barely recognize it. The heavy seas setting upon this shore during the late almost continuous succession of north-westerly gales had washed away the old beach, and thrown ridge upon ridge of shingle higher up the embankment, bringing the spot where our tent stood some yards nearer to the water's edge. We could nowhere find the horns; they must either have been washed away or buried beneath the confused heaps of shingle and huge hummocks of stranded ice.

At one p.m., after rounding Cape M'Bain, we hauled the boat up on a hard shingly beach, on the north side of Clark Bay, about half-way up, and pitched the tent on a fine, dry part of the ridge, on the margin of a frozen lake. Saw several gulls sitting on the beach; and just as I landed a solitary raven (*Corvus corax*), hovering overhead to reconnoitre our proceedings, fell a victim to his curiosity. I fired at him, and he fell dead upon the surface of the frozen lake. This bay appears to be a favourite retreat of the ravens; we saw several on our last visit here, but none elsewhere. At three p.m. we had our usual warm mess for dinner, and opened the last gutta-percha case of biscuit. Three of our party having eaten rather too freely of the bear's liver for supper last night, complained to-day of violent headache, which readily yielded to a smart cathartic dose of medicine.

At five p.m. I left the encampment, accompanied by one of my party, on a searching excursion over the ridges round the bay, to the summit of Cape Bowden, a distance of about six miles from the tent. Our course lay over a succession of ridges, and through ravines filled with deep snow, in many places above the knees at every step we set, and in the snow-drifts crossing some of the deep hollows even up to the waist. We had to climb one very steep hill, separated from Cape Bowden by a deep, saddle-like depression, nearly filled by a frozen lake. We rapidly descended to this, but had another toilsome ascent up the steep acclivities of the Cape; and on reaching the summit had to walk a mile further over deep snow before I found the *Rescue's* cairn, which stands on the southern extremity of the ridge. We reached the spot at seven p.m. I drew from beneath the pile of stones a broken common green quart bottle, containing a gutta-percha case, enclosing the usual printed notice on yellow paper left by the searching parties from the *Lady Franklin* and *Sophia.* I tore a leaf from my memorandum book, and wrote on it a record of my visit, which I put in, and replaced the bottle in the cairn. Having taken a rough sketch of the coast, extending from Point Bowden to Cape Spencer, the whole outline of which appeared displayed beneath as on a map from this elevation, I commenced my return, and on reaching the extreme craggy north point of the ridge, I took another sketch of our encampment on the other side of Clark Bay, with Cape Grinnell and the headlands seen jutting out beyond it to the north. The spot on which I stood was a rugged crag, overhanging Wellington Channel; the chasm or deep gorge, which cleft the crag in two, forming a steep and precipitous descent to the beach below, was in part treacherously arched over with a frail crust of snow, rendering it a dangerous place to approach in a thick snow-drift, as one false step would hurl the wanderer headlong into the frowning gulf below. The brown, weather-worn surface of the limestone strata was so arranged in horizontal layers on either side as to resemble reams of brown paper piled one above another more than anything else; as these vertical sections, on which the snow could find no resting-place, peered from beneath its otherwise universal covering of the land. In the valley beneath lay the still frozen surface of the lake. Looking up channel the northern horizon presented a very remarkable tint of the deepest indigo blue—a peculiar tint, I do not recollect ever having seen before, and bounding it like a narrow band or streak, the sky elsewhere being overcast all round, with the exception of a wild glare of light which gleamed through the black canopy shrouding Cornwallis Land on the opposite shore. I heard the lively note of the snow-bunting, the only indication of life around us in this still and desolate solitude. We saw neither bird nor beast else throughout the whole of our excursion. Occasionally a track of the fox or hare met the eye, and we saw the footprints of the ptarmigan (*Tetrao lagopus*) on the acclivity of Cape Bowden.

After descending from the crag into the valley to the lake beneath, we toiled up the steep face of the ridge on the other side, not a little jaded and fatigued with the rough and rugged outward journey, and the agreeable prospect before us of a return over the same course, now with monotony instead of novelty for companionship. It was now eight p.m., and we reached the tent at ten p.m. Thermometer down to 21°. Night foggy, with light airs.

Tuesday, 7th.—The wind this morning suddenly shifted round to the south-west, accompanied by a fall of snow, which, with a strong breeze blowing, confined us to the tent until about five p.m., when the weather cleared up, but the wind being against our going down channel, together with some heavy streams of bay ice in the offing, brought over from the opposite shore by the shift of wind, delayed our departure to-day.

At the time of setting the first watch for the night, the moon appeared in a bright crescent form, shining forth through an opening in some light, fleecy clouds, which were passing across the clear, blue, ethereal sky; the evening star was peeping over the ridge at the back of the tent, twinkling with unusual brightness, just above a faint red streak of light which skirted the horizon; and here and there a star of the first magnitude was just becoming visible in the zenith and the western portion of the heavens. The thermometer had fallen to 24°.

Wednesday, 8th.—This is the first fine day that we have really had since we left the ship; the sun, which for the last three weeks has been an entire stranger to us, now shone forth from a clear blue sky. When I registered the thermometer, however, at six o'clock this morning, it was as low as twenty degrees below the freezing point, having fallen no less than twelve degrees during the night—from 24° to 12°. The maximum during our voyage of three weeks was only 31°, minimum 12°, and the mean 21°, never having at any time risen above the freezing point. The mean of eight days, taken with the aneroid before it was damaged, was 29° 54′. It was bitterly cold within the tent, my south-wester, mitts, and Esquimaux boots were hard frozen under my head, where they had formed a substitute for a pillow.

After breakfast we built our last cairn on the spot of our last bivouac, and buried beneath it a tin cylinder containing the following record of our proceedings:—

MEMO.—A boat expedition up Wellington Channel in search of Sir John Franklin. Left her Majesty's ship *North Star* at Erebus and Terror Bay, Beechey Island, on Thursday morning the 19th of August, and after a close examination of Baring Bay by sledging round its shores on the snow, without finding any opening to the eastward, on returning down channel searched every bay, inlet, and headland along the coast without discovering any traces of the lost ships. Encamped here on Monday, September 6th, and the boat is now launching to return to the ship. The weather throughout the whole of this time has been most tempestuous—continued gales of wind, accompanied by thick weather and a short, broken sea with a heavy swell, very dangerous for boats. The thermometer, which has never been above the freezing point, fell last night twelve degrees, from 24° to 12° Fahr. The young ice formed in the bay, and the whole of the land is enveloped in a white mantle of snow. But few animals have been seen, vegetation being very scanty. Traces of the musk ox, however, and its horns were found, and three hares seen in this bay. On Saturday last I shot a large bear on the south side of Griffin Bay.

R. McCORMICK, Officer Commanding Party.

Wednesday, September 8th, 1852.

Having struck the tent and stowed the boat, we launched her at 10.30 a.m. and made sail with a fresh and fair breeze round Cape Bowden, outside of which there was still a short broken sea in the channel; but we carried on through it without taking in a reef. Reached Cape Spencer at four p.m., after a fine passage of five hours and a half, under sail the whole way. Here we hauled the boat into a small creek between the shingle ridges, and lighting a fire on the bank of shingle, commenced cooking our dinner, when a boat under sail, and standing towards us, hove in sight, coming round the point of the shingle spit which divides Erebus and Terror Bay from Union Bay, and on which the graves are situated. On reaching us we found that it was the second gig, in charge of the second master, with provisions to be left *en cache* at Cape Osborn; but she was far too late to have the slightest chance of accomplishing this object. In my own mind I doubted much her reaching even our last place of encampment, which we left this morning.

The news we obtained from her of the arrival of Lady Franklin's vessel, the *Isabel*, from England during our absence, having only sailed again this morning on her homeward voyage, was quite unexpected.

I winged a young silvery gull here (*Larus argentatus*), which I took on board with me alive.

As soon as we had finished our dinner we shoved off, and instead of going round Beechey Island made sail across Union Bay (the winter's floe having cleared out during our absence). The moon shone brightly forth just above the cairn on the summit of Cape Spencer. Rounding the point of the spit, on opening the *North Star*, she hoisted her colours, and we ran up ours; the bugle sounded on board, which I answered by firing off my gun.

At 8.30 p.m. ran alongside, when I had the pleasing gratification of finding letters for myself from home, sent me by my friend, Mr. Barrow, of the Admiralty, together with piles of newspapers brought out for the squadron.

The following is a list of the names of the crew, selected from ten volunteers who offered to accompany me in H.M. boat *Forlorn Hope*.

Thomas Rands,	æt.	30,	Ice-quartermaster.
Edward Millikin,	,,	25,	Able seaman.
James Nugent,	,,	27,	Able seaman.
Eleazer J. Clark,	,,	39,	Royal Marines.
George Burns,	,,	25,	Able seaman.
John Frost,	,,	23,	Able seaman.

Of the cool, steady, praiseworthy conduct of all, the unanimous good feeling shown towards each other, and respectful attention to myself on all occasions, I cannot speak too highly. They deserve my warmest approbation; and I trust that their meritorious services will not pass unnoticed. Thomas Rands I found a most able and efficient petty offcer. He also gave universal satisfaction in serving out our daily rations, which I committed to his charge.

R. M'CORMICK, R.N.

APPENDIX.

LIST OF GAME KILLED BY R. McCORMICK, R.N.

		Number
Polar bear	*Ursus maritimus*	1
Arctic fox	*Canis lagopus*	2
Arctic hare	*Lepus borealis*	6
Lemming	*Georychus lemmus*	1
Raven	*Corvus corax*	4
Ptarmigan	*Tetrao lagopus*	2
Greenland finch	*Fringilla greenlaudiac*	2
Sandpiper	*Tringa maritima*	4
Little auk	*Alca alle*	4
Dovekie	*Uria grylle*	10
Loom	*Uria brunnichii*	2
Red-throated diver	*Colymbus septentrionalis* (young)	1
Tern	*Sterna arctica*	1
Fulmar petrel	*Procellaria glacialis*	2
Ivory gull	*Larus eberneus*	2
Silvery gull	*Larus argentatus*	2
Glaucous gull	*Larus glaucus*	1
Eider duck	*Anas mollissima*	9
Pintail duck	*Anas caudacuta*	3
King duck	*Anas spectabilis*	1
	Total	60

DIMENSIONS OF THE POLAR BEAR (Male) *shot September 4th, 1852, in Griffin Bay, Wellington Channel.*

	Feet.	Inches.
Length	7	6
Greatest circumference of body	5	6
" " head	2	8
" " neck	3	3

	Feet.	Inches.
Length of head	1	6
„ of fore leg (from shoulder joint) . . .	3	2
Circumference of fore leg	2	2
Length of hind leg (from hip joint	3	0
Circumference of hind leg	2	0
Length of fore paw	1	0
Circumference of fore paw	1	9
Length of hind paw	1	9
Circumference of hind paw	1	9
Estimated weight	1000 lbs.	

CONCLUDING REMARKS

ON THE

SEARCH FOR SIR JOHN FRANKLIN,

The probable Position of the "Erebus" and "Terror," and Fate of their Crews.

MY experience during the late voyage and winter, passed on the very same spot where Franklin spent his, and where all traces of him cease, have most decidedly confirmed me in the opinion I had ventured to express in my plans of search some five years ago,—viz., that the missing expedition passed up the Wellington Channel into the Polar Sea, and was to be sought amongst the archipelago of islands and drifting packs of ice with which that sea is most unquestionably encumbered, and where the search should be made with efficient well-equipped boats, adapted for encountering the packs of ice, strong currents, and dangerous intricacies, inseparable from such a navigation, promising nought else but destruction to ships. From boats alone could any hope be entertained of a rescue of our gallant countrymen, ere they fell victims to the combined effects of frost and famine,—for in these two expressive words all their privations may probably be summed up,—and, if too late to save them, of discovering any traces they may have left behind them.

At that early period of the search I believe I stood alone in this opinion. The general impression was that the ships had been arrested in the ice to the southward and westward of Melville Island; consequently, the main efforts for carrying on the search took that direction. There are few perhaps who will now dispute my views, or their originality, which the Parliamentary records have secured.

My reasons for coming to the conclusion I then did need not be recapitulated here, they having been fully explained in my plans submitted at the time, and, subsequently, in the year 1850-52, accompanied by the first proposal made for attempting the search in so high a latitude in an open boat, which I volunteered to conduct. This plan obtained the warm support of the Hydrographer, Rear-Admiral Sir Francis Beaufort, and of Rear-Admiral Sir Edward Parry (with whom I made my first voyage towards the North Pole), who recommended my employment in very favourable terms in their reports of approval annexed to my plans laid before Parliament.

I was at last sent out in the *North Star*, but the position I was necessarily placed in in that ship was not such as to enable me to act in the noble cause in the way I had hoped, and, being somewhat anomalous, renders it incumbent on me to be careful that my share in the search is not left open to misconception. Here I may, therefore, be permitted to draw attention to the fact that, could I at once have proceeded up the Wellington Channel on the first arrival of the *North Star* at Beechey Island, on the 8th of August, 1852, with my boat's crew of volunteers, instead of being detained until the 19th of the same month,—by which delay we lost the last eleven fine days of the season, and best portion of it in which boating operations can be carried on in those seas, Wellington Channel being as open as the Atlantic, as far as the eye could reach from the summit of Beechey Island, which, with Cape Riley, I ascended on the day of my arrival; the season an unusually open one, with little or no ice, and the wind blowing from the southward and eastward, fresh and fair,—there was nothing to have prevented us from doubling Cape Sir John Franklin, and proceeding round by Jones Sound into Baffin Bay, before the north-westerly gales set in, which at a later period we met with; these winds would have proved fair for our return down Jones Sound, sheltered under the lee of the land, round by Lancaster Sound and Barrow Strait to Beechey Island, thus completing the circumnavigation of North Devon, and an entire examination of its shores. Subsequent events have proved that all this might have been accomplished in the season.

When we were enabled to get away from the ship winter had already, the day before, set in. After an absence of three weeks' exposure to a succession of north-westerly gales, and, altogether, the most boisterous weather that I ever before experienced, as described in the preceding narrative; I, however, had the satisfaction of setting the Baring Bay question at rest, viz., that there is no communication whatever between that bay and Jones Sound.

After my return I wrote a letter to the commander of the expedition early in the spring, offering to explore Smith Sound into the Polar Ocean as far as the season would permit of, if I was given the command of the *Mary* yacht, a decked boat of twelve tons, cutter-rigged, and well adapted for such a service, as, in addition to the greater quantity of provisions and stores which she would stow for a prolonged search, she would also possess the advantage of greater safety in a sea that might endanger an open boat, more especially if deeply laden, as the *Forlorn Hope* was. My former boat's crew having volunteered to accompany me again, and cheerfully expressed their willingness to follow me wherever I led them, it was my intention to have brought the *Mary* across the Atlantic home, after completing provisions and fuel at some one of the depôts at Pond's Bay, or the southern shores of Lancaster Sound, instead of risking her getting beset for the winter in the heavy packs with which Barrow Strait was filled this season.

My object in the voyage up Smith's Sound was to have made as near an approach to the Pole as the state of the ice would have permitted. I believe that if ever the North Pole is reached, it will be on the meridian of Smith Sound. (It will be seen in the sequel that I have had reasons for changing my opinion.)

I may here offer a few suggestions on the probable fate of the missing ships and their crews, having myself entertained sanguine hopes of discovering some traces of them in the higher latitudes, which it was my intention, if possible, to have reached had the command of the *Mary* been given me. This, however, was declined by the commodore, and in the anwer which I received from him to my offer, dated on board the *Assistance*, the 26th of July, 1853, the reason assigned was that "Nothing now remains undone in that vicinity." Every hope of making myself further useful in the cause being now at an end, I had no other alternative left me than to return home in the *Phœnix*, having done all that it was in my power to do.

There are several ways by which a ship may be destroyed—by fire, by foundering, by collision with ice, or by being driven on shore. Either of the first two casualties might easily enough happen to a single ship; but as it is in the highest degree improbable that two ships should together share the same fate, these two modes of accounting for the loss of the Polar ships may at once be disposed of. The third, by collision with ice, carries with it a greater amount of probability. Even this, however, in the case of the *Erebus* and *Terror* seems to me a very unlikely catastrophe to have happened to two ships so strongly built, and so well additionally fortified by the stoutest doubling, as those ships were, rendering them capable of resisting an amount of pressure from ice truly astonishing, as I can, from my own personal observation, vouch for, having seen them beset in the immense packs of ice in the Antarctic seas, consisting of floes mostly of great thickness and density, the latter quality being greatly increased from the temperature never rising above the freezing point within the Antarctic circle even at midsummer, consequently exerting no thawing influence on those vast fields of ice, which, when put in motion during the agitation of the great southern ocean by heavy gales, I have often seen the strength of the *Erebus* most severely tested between huge dense masses of blue ice, violently grinding past her sides, tearing and rolling up her stout copper sheathing like so much ladies' curl-paper, whilst every beam and timber in her have been creaking and groaning, and the rudder almost wrung from its fastenings. Ice with which the floes and packs within the bays and straits of the Arctic circle can no more be compared, than the ice on the surface of the Serpentine can with the floes of Melville Bay. The only ice I have ever seen in the north at all to be compared with the southern packs occurs in the Spitzbergen seas.

I have entered more fully upon the effects of ice than I should otherwise have done, in consequence of having frequently heard the loss of the *Breadalbane* hired transport cited as an example of the loss of Sir John Franklin's ships, many persons jumping at once to the conclusion that the latter must have been crushed and engulfed in the same summary way as the unlucky transport was. The two cases, however, are widely different. The *Breadalbane* was known to be an old vessel, which the owners had not sufficiently doubled or strengthened to enable her to resist even a moderate degree of pressure from two contending floes; the consequence was they went through her bottom, and she disappeared beneath them within a quarter of an hour from the time she was first caught in the "nip," as I was a witness to myself from the deck of the *Phœnix*, which ship was in the same "nip."

The American whaler, *M'Lellen*, lost in Melville Bay in the season of 1852, is another instance brought forward in support of this opinion; but she, also, was an old, worn-out ship, and her timbers very defective where the floe-edge caught her side and stove her in. This I saw myself, as I went on board of her at the time, she having become a wreck immediately under the bows of the *North Star*, carrying away that vessel's cathead. But to draw any comparison between those two vessels and the *Erebus* and *Terror*, would be like comparing the cracking a hollow filbert with the hardest ivory nut.

Much has been said about the ships having been forced out of Erebus and Terror Bay, and of their having left indications behind them of a hurried departure. On what grounds these surmises have been founded it would be somewhat difficult to divine. I passed a whole year in that bay, and whatever may be said to the contrary, I believe it to be utterly impossible that any vessel could be driven out of it after having once been frozen in; a more safe bay for wintering in does not exist along the whole line of coast. Its very fault lies in its security, the difficulty in getting out again when once within it, as the bay-floe rarely breaks up before the end of August or beginning of September. The *North Star* getting on shore there had nothing whatever to do with the bay, and was an event in no way calculated to compromise its character for safety. The spot where the *Erebus* and *Terror* lay was evidently near its western extremity, in the curve of the bay formed by the shingle ridge, extending out from Beechey Island, on which "the graves" are situated; the close vicinity of the magnetic observatory, the armourer's forge, the washing-place at the water-course, and the small garden not much farther off, with the cairn above it—all combined to point out this as the winter quarters of the ships, and a more secure one could not well have been fixed upon. In fact it was the only position in this bay in which a ship would be altogether secure from being driven on shore by any sudden ingress of ice in the autumn, before the winter's floe was firmly formed, and, as such, could not fail to have been selected by one of Sir John Franklin's judgment and experience. I saw nothing whatever in support of the notion that the departure of the ships was a hurried one, but much to convince me that Franklin and those with him had not idly passed their winter here, to which the sites of tents in various directions, sledge-tracks, and everything else bore ample testimony.

Further, I am of opinion that sledging-parties from his ships had been up the Wellington Channel, and reasoning upon what I know may be accomplished, even in mid-winter, where energy exists, as in such men as Franklin and my lamented friend, that soul of enterprise, the noble-minded Bellôt, these sledge journeys were very probably extended beyond Cape Lady Franklin—even to the portal of the Polar Ocean. Their tracks round Cape Spencer in the direction of Cape Bowden, clearly point out the course they had in view; here no inducement could be held out to the sportsman to tarry, there is not even sufficient game for a single gun, far less to render it an eligible spot for pitching a tent as a mere shooting-station.

The swampy flat, intersected by small lakes and water-courses, in the vicinity of Caswall Tower, is the only spot where the very few straggling wild-fowl that alight in this barren limestone region, on their way north, are to be met with; and here I have followed Franklin's sledge tracks over the low shingle ridges in the direction of the tower, which was doubtless their shooting-station. The sledges must have passed in the summer season when the soil was plastic enough to leave impressions of their tracks behind them. Caswall Tower is an isolated precipitous mount between three and four hundred feet in height, rising from a plain at the head of Radstock Bay and Gascoigne Cove, which I ascended, but found nothing whatever on its bare flat top, save a solitary lemming, which I captured. At its base are several circular ancient Esquimaux encampments, within which the wild flowers flourish more luxuriantly than in any other spot I met with. The distance is about ten miles from the ship.

The greatest mystery of all is, that of no record having been left of their sojourn or departure; so sanguine was I for a time that something might turn up to reward a diligent and persevering search, that I did not rest until I had closely examined every foot of ground for miles around; ascending and descending every hill and ravine around the bay, and rambling over the mountain limestone table-land, far inland, till there was not a rock or ravine on the land, or hummock of ice on the floe, within a circuit of many miles, that was not as familiar to me as "household words."

From my own experience, throughout a somewhat more severe winter, perhaps, than ordinary, I believe that sledge-travelling may be continued during an Arctic winter, without much risk or danger being incurred from the lowest temperature; provided care is taken to erect a snow-hut, or, in cases of emergency when no time is to be lost, to cut a deep trench in the snow in time to secure shelter from an approaching gale and snow-drift. It must be kept in mind, that the same degree of cold which can be borne without inconvenience in a calm cannot be faced without severe frost-bites in a strong breeze of wind.

In thus recording my opinion of the practicability of sledge-travelling in the winter season, I have the testimony of those enterprising Arctic travellers, Kennedy and Bellôt, in my favour, who, during the *Prince Albert's* voyage, practised it most successfully in mid-winter. I also have had opportunities of fairly testing the effects of a very low temperature on my own person on more than one occasion. My customary walk throughout the winter, whatever the state of the weather might be, was round Beechey Island, a distance of about six miles. This I accomplished once when the thermometer was 54° below zero on the floe, and to that low temperature I was exposed for two hours, without feeling any inconvenience from it, but there was little or no wind at the time.

On another occasion I passed a whole day and night without food or shelter, beyond what the snow-drift afforded, about seven miles from the ship, having been overtaken by a dense fog on the open plain when returning from an excursion to Caswall Tower, accompanied by my friend Mr. Toms, of the *North Star*, and Erebus and Terror, my two Esquimaux dogs. When overwhelmed by the darkness of night blending with the fog, and a gale approaching, we cut with a hunting-knife a trench in the snow-clad plain, about two feet deep, and in this truly Arctic bivouac (at all times to be found) we, with our canine friends, passed the night, without a tent or other clothing than our usual walking-dress.

The gale which swept over us soon forming a white coverlet of snow-drift, protected us from the blast, less than an hour's exposure to the inclemency and intensity of cold of which would inevitably have ended in our destruction; not even the dogs would have survived it. The thermometer that night fell to 32° below zero, or 64° below the freezing-point. The fog clearing off sufficiently to make out the land, about four o'clock in the morning we started again, and reached the ship between six and seven a.m., without having incurred even a frost-bite, and after an ablution and breakfast, felt as fresh as ever.

Sketched by R. M'Cormick, R.N.

Caswall Tower, from Beechey Island.

Page 168.

I am, therefore, led to the conclusion, that Sir John Franklin's travelling-parties may have commenced their journeys up the Wellington Channel, with the first appearance of the sun above the horizon early in the month of February, and after the discovery that the strait between the Franklin Capes in the Queen's Channel opened into a Polar ocean, started with his ships as soon as the bay ice broke up, most probably about the first week in September; and if he had an open season would, with the aid of his screw-propellers, run up the Wellington Channel within the space of eight-and-forty hours. Then, probably tempted by the broad expanse of open water to the northward, or at any rate absence of land to obstruct his progress in that direction, he might reach a very high latitude, and gain a good offing of the Parry Islands, before he shaped a south-westerly course for Behring's Strait. As the season for navigation remaining after the 1st of September, would be, however, necessarily a very short one, he was probably overtaken by winter, perhaps some six or seven hundred miles from Cape Lady Franklin, in a high latitude, and possibly well to the westward.

Having thus attempted to follow up the track of the unfortunate ship so far by something like inductive reasoning, founded on inferences drawn from a knowledge of the object they had in view, and the most probable events and incidents likely to beset them in their path to mar its attainment, we now enter upon a field of speculation, wide enough indeed to fill a volume of itself. Having already extended these remarks to a greater length than I had intended, I will wind them up with a few words on the conclusion I have come to in my own mind, as to the fate of our gallant countrymen. Speculative as any opinion upon this subject, I am aware, must necessarily be, I have not arrived at mine either prematurely or hastily. No one but those who may have near relatives in the expedition, can possibly have felt deeper interest in this hapless search from first to last than I have, unless it is my friend Mr. Barrow, whose untiring exertions and devotion in this noble cause stand unequalled. Various associations combined to enlist my own sympathies in this search. They were my old ships, and engaged in a field of discovery to which I have long been ardently devoted, and in which my thoughts have been centred from my earliest youth, in addition to which there were those on board of both ships who were well known to me.

My own impression is, that on the closing in of their second winter, the ships were either driven into some inlet, where they may have been blocked up ever since by the Polar pack, as happened to the *Investigator* in Mercy Bay; or that they may have been driven on shore by the strong currents which set from the north-westward, when helplessly beset in the pack, drifting about in the narrow straits which separate one island from another in this Arctic archipelago.

They may, possibly, have reached even as far west as that large tract of land whose mountainous and lofty granitic peaks were seen by the *Herald*, thus barring their further progress westward. But, under any of these circumstances, it does not follow that the lives of those on board would be necessarily involved in immediate destruction, even were the ships stranded on some shore. They would, in all probability, be able to save the greater part of their provisions and stores (as Sir Edward Parry did in the loss of the *Fury*, on Fury Beach; and which, years afterwards, proved the happy means of preserving the lives of Sir John Ross and his party). They might build huts and supply themselves with fuel from the wreck, and linger out an existence as long as their resources lasted. But here, however reluctantly, I must at the same time acknowledge that there would be but little prospect of adding much to these in the region in which their disaster would be likely to happen. In proof of this, I have only to add that had I lost my boat and the provisions when up the Wellington Channel, my boat's crew and myself could not have existed—although numbering only seven—on the produce of our guns, for one month; and I had two or three good shots in my party, besides being myself an old sportsman, and rarely threw away a shot without obtaining something for it. Wild-fowl, doubtless, migrate to the very pole itself to rear their young, but this occupies only a short period of the season; and the supplies to be obtained from such an uncertain source would be inadequate even for present wants, far less so to form a winter's store for a ship's company.

Sad as the reflection must be, it is in vain to deny that the time has arrived when, indeed, it is "hoping against hope," and which suggested to me the name of "Forlorn Hope" for my boat. Nearly nine years have now elapsed since our countrymen left these shores; and although I have been to the last one of the most sanguine in my hopes, I cannot help feeling now that traces of their fate, is all, unhappily, I have too much reason to fear, that remains to be discovered of them. But even this, in my opinion, will never be accomplished by ships. Nought else than the disastrous fate of the gallant Franklin and his followers can be possibly anticipated as the result of any attempt made by ships.

R. McCormick, R.N.

SUGGESTIONS

FOR THE

PRESERVATION OF HEALTH IN POLAR CLIMES.

HAVING had under consideration the best means of escaping scurvy, and preserving health in the Arctic regions, I deem it my duty to submit the following brief remarks for the use of future voyagers. In so doing, I shall confine myself wholly to the results of my own experience during a period of some years passed in the higher latitudes, feeling confident that attention to the precepts here inculcated will secure for those who may follow me as successful an exemption from scurvy and sickness as have crowned my own efforts, by a rigid adherence to them.

In the first place, I would unhesitatingly recommend the *entire exclusion* of all kinds of salted meats from the diet, convinced as I am, from long experience and close attention to the effects of such food, that it proves, through its indigestibility and deficient nutriment, properties injurious to the system, and deteriorating the condition of the circulating fluids and secretions generally,—inducing a debilitated habit of body, favourable to the production of scurvy, under circumstances of privation and exposure, and other exciting influences, calculated to call it into action. In fact, it is my belief that the origin of every case of scurvy may be fairly traced to the use of salted meats.

In the present age of inventions and improvements, there can be no lack of substitutes, and excellent ones too, for the hard salt beef and pork, and the whole category of dried tongues, hams, &c., which constituted the sea stock of bygone years, when every ship in a long voyage, as in Anson's time, lost great numbers of the crew.

Now, we have preserved meats, poultry, soups, pemmican, and fresh bacon of all kinds. The latter article, which was supplied for the first time to the expedition now out, especially that preserved in tins for the use of travelling-parties, proved the most valuable addition of all to the scale of Arctic victualling, its freshness and mildness rendering it easy of digestion, and its fatty quality rendering it highly nütritious by affording a large supply of carbonaceous material to make up for the constant waste occasioned by the increased exhalation of carbon which accompanies the activity of the respiration in very low temperatures of the atmosphere.

The various kinds of vegetables, when carefully selected and preserved, are quite equal to the fresh ones; more especially the preserved potato, carrot, parsnip, turnip, and peas; and I cannot speak too highly of those bottled fruits, as the damson, greengage, currant and raspberry, gooseberry, and that perhaps best of all anti-scorbutic fruit, the cranberry, which is quite equal to the lime-juice in its valuable properties: all these fruits are quite as good as when first gathered.

Dried fruits—apples, figs, prunes, raisins, and almonds, &c., are all objectionable.

The best diluents are tea, coffee, and chocolate, more especially the patent chocolate which the travelling-parties were supplied with in the last expedition. Of spirits and wines, the less taken the better; good sound malt liquors are preferable in all respects, combining, as they do, a nutritive with a stimulating property.

On the subject of clothing, I have only to observe that I found the Government pilot-cloth suit, with a "sou'-wester," the most generally useful in summer or winter; but for boating or sledging, in severe weather, I know of nothing equal to the Esquimaux seal-skin dress and fur boots. A common blanket bag I have always found far more comfortable than a felt one for sleeping in, when away travelling, with a buffalo robe beneath it.

Of medical treatment, little is required. The bracing effects of a low, dry temperature, and the absence of all moisture in the atmosphere for a large portion of the year, so that not a cloud can form in the clear blue sky, render catarrhal and other affections resulting from atmospheric transitions of rare occurrence.

During the dark and monotonous season of winter, active exercise in the open air, on the floe or on the land, is the very best preservative of health, aided by proper attention to diet; the mind being at the same time engaged in rational occupations, reading, writing, sketching, or whatever may be the bent of individual taste.

When sufficient exercise is not taken, and the diet has been too full and liberal, a congestive state of the internal organs is often the result, attended with a drowsiness during the day and broken rest at night. The best remedy I have found in such cases is a six-grain dose of calomel, and, to allay the disposition to watchfulness, about a scruple of the compouud ipecacuanha or Dover's powder, given at bed-time. Loss of appetite, from want of tone and energy in the digestive organs, sometimes follows the effects of a long and tedious winter in some constitutions. A wineglassful of quinine wine, given twice a day, is the most efficacious remedy in these cases; it is best prepared by dissolving about a scruple of quinine, with the same quantity of citric acid, in a wineglassful of water, and then adding it to a bottle of wine, either port or sherry, as may best suit the occasion.

In conclusion, I have only to add, in confirmation of these views, that in three voyages which I have made to the Polar regions—two to the north and one to the south, the latter of four years' duration,—embracing every possible transition of climate and exposure, I have never lost a single life, or even had a case of serious sickness or scurvy throughout a period of Polar service falling little short of seven years.

R. McCormick, R.N.

CORRESPONDENCE.

To Captain Sir E. BELCHER, C.B., H.M.S. *Assistance.*

Her Majesty's Ship *North Star*,
Erebus and Terror Bay,
3rd March, 1853.

SIR,—I have the honour to transmit to you a narrative of my boat expedition up Wellington Channel, and sledge journey round Baring Bay, in search of Sir John Franklin.

Having left the ship on the morning of the 19th of August, and returned on board again on the night of the 8th of September last, after an absence of three weeks, during the whole of which time the weather was most unfavourable for boat service, having been tempestuous and overcast in the extreme,—a succession of north-westerly gales, which, with strong currents, rendered the navigation of this channel a very dangerous one for boats, and not a place of shelter between the last bay and Baring Bay.

After a week passed in a most careful search of Baring Bay all round, and ascending the inland ridges of hills, I neither found an opening to the eastward or a surface practicable for sledging over inland; the whole forming a succession of steep ridges, with intervening ravines filled with snow, and running parallel with the top of the bay.

There was no indication whatever of open water in the

vicinity; the gulls and other sea-fowl never shaped their course to the eastward.

Therefore in all probability Jones Sound, instead of continuing its course to the westward from Baffin Bay, soon trends round to the north-west. On my return down channel I carefully examined every headland and bay, unhappily without finding the slightest trace of the missing ships.

Five of these bays, and several of the most prominent headlands between Point Bowden and Cape Osborn, not laid down in the charts, I have availed myself of the usual privilege of explorers, and given names to them.

My party returned on board in good health; and I have great satisfaction in bearing testimony to their exceeding good conduct, and they having volunteered to accompany me again in the spring search, I have herewith to submit for your consideration my purposed plan for carrying out that search.

In your letter of the 13th of August last I was told that the *Assistance* and *Pioneer* would complete the search of the Wellington Channel, and that my course must be to the eastward of this meridian. Sledging, therefore, will be entirely out of the question, as Lancaster's Sound opens too early to permit of travelling over the ice to any distance and back again.

The boat, however, which I had last (and we have none better adapted on board) is wholly inadequate for so long a voyage as the one contemplated, viz. the exploration of Jones and Smith Sounds, more especially as since your departure Commander Inglefield, in the *Isabel*, has been so far up both these sounds as to render it very improbable that a boat, stowing barely a month's provisions, could remain out sufficiently long to enable her to accomplish anything beyond what he has already done.

The plan, therefore, I have to propose is, that the

Mary yacht, left by Admiral Sir John Ross, and now lying here useless, should be placed at my disposal, with two additional hands, and provisioned for three months, with a gutta percha boat (left here by the *Prince Albert*) for hauling over the ice, should the floes in the sounds not have broken up. To start immediately after the return of the party, conveying your authority so to do, and by which time the navigation in Barrow Strait will most probably be open.

I am, &c.,

R. McCormick, R.N.

P.S. The departure of the sledge-parties for the rendezvous depôts, being a month earlier than anticipated, a series of sketches, comprising the headlands and bays between Beechey Island and Point Hogarth, Baring Bay (which I had taken for the purpose of illustrating a track-chart on which they are laid down from compass bearings), not being finished, I must reserve for a future opportunity.—R. McC.

The Secretary of the Admiralty to Surgeon McCormick, R.N.

Admiralty, 13th of October, 1853.

Sir,—I am commanded by my Lords Commissioners of the Admiralty to acknowledge the receipt of your narrative of an expedition under your orders in a boat of H.M. Discovery Ship *North Star*, up the eastern shore of Wellington Channel and round Baring Bay, for the purpose of discovering traces of Sir John Franklin's expedition.

My Lords approve of your exertions on this occasion, and of the conduct of your boat's crew on a service incurring both risk and hardship, and are satisfied with the efforts you made in determining the important question as to there being any connection between Baring Bay and Jones Sound.

I am, &c.,

(Signed) W. A. B. HAMILTON.

PART IV.

AUTOBIOGRAPHY.

Vincent Brooks, Day & Son, Lith.

SURGEON ROBERT McCORMICK, R.N.

FATHER OF THE AUTHOR.

TO FACE TITLE PAGE OF AUTOBIOGRAPHY. VOL. II.

Quercus McCormickii, or Tartary Oak.

AUTOBIOGRAPHY.

CHAPTER I.

Early life—Career as medical student—Entering the navy.

My father, the late Robert McCormick, a surgeon in the Royal Navy, was born under the same old homestead in which his father, also named Robert, first saw the light—Ballyreagh, in the county of Tyrone, Ireland, which with its lands has been for generations in the possession of the family. The estate is situated between Cook's Town and Dungannon, near Lough Neagh. My father entered the Royal Navy in the year 1795; his first appointment to H.M.S. *Powerful*, of eighty-four guns, bears the date of the 17th of July, from which ship he was promoted to the *Inspector* sloop-of-war; and after a career of much war service, lost his life in the shipwreck of H.M.S. *Defence*, of seventy-four guns, on the Christmas Eve of the year 1811, when convoying a disabled ship, the *St. George* of ninety-eight guns, bearing the flag of the late Rear-Admiral Reynolds, homeward-bound with the Baltic fleet. Both ships were wrecked in a dreadful hurricane within a mile or two of each other, on a reef off the coast of Jutland, and with the exception of a few of the seamen from each ship thrown on shore clinging to the spars, their crews, numbering some 1300 and upwards of officers and men, perished in this melancholy catastrophe.

I was born at the commencement of the century, on the 22nd of July, 1800, in the village of Runham, near Great Yarmouth, in the county of Norfolk, where I passed the first six years of my life, not very far from the birth-place of the immortal Nelson, England's greatest admiral, who, like myself, in early life evinced a strong bias for Polar discovery, having accompanied Captain Phipps (afterwards Lord Mulgrave) in his attempt to reach the North Pole in the latter part of the last century, as I did in the early portion of the present one under the command of Parry.

So attached was I to a rural life, that in subsequent years, until I had completed my second decade in life's journey, I was indeed a very frequent sojourner in my native village. Notwithstanding that, destiny had decreed the best part of my life for the future was to be passed on a widely different element, the boisterous ocean. My taste for natural history developed itself at an early age, beginning with the collection of the nests and eggs of the birds of the surrounding district.

Fronting my birth-place a large tract of marsh and meadow land extended over what once formed an estuary between the now port of Yarmouth and at present inland city of Norwich. Bounding the horizon to the south, the Suffolk uplands are plainly visible, and two very striking clumps of trees opposite indicate the site of the ancient Roman ruin of Burgh Castle. Through this extent of marsh-land the river Yarè winds its course, with here and there the picturesque sails of a water-mill and its adjacent thatched house, or the brown-tanned sails of a Norfolk wherry, passing and repassing with a freight of the produce of the farm, consisting of grain, &c., for the markets.

These meadows then abounded in the spring of the year with plover, snipe, and other water-fowl, and were the favourite breeding-place of the lapwing (*Vanellus cristatus*), where my early morning excursions usually led to a discovery of their nests and beautifully mottled eggs, having an olive-green ground, blotched over with rich brown spots four in number, most symmetrically placed with the smaller ends converging downwards in a superficial depression, formed generally on the margin of some dried-up stream, where the grass had become withered and yellow. Over this lowly nest the excited parent birds, in their anxiety for their treasure, wheel round the intruder's head with their large slowly-flapping wings in a circular flight, repeating their pleasing yet plaintive and somewhat melancholy cry of "Pewit, pewit," and endeavouring with great tact to lead him away from the tabooed spot. Whilst the Jack snipe, soaring on high, suddenly descends with the rapidity of an arrow shot from the bow, and with a loud whirring sound as he approaches in his descent the nest so carefully concealed within a tuft of long grass, in which small "wigwam" the partner of his domestic cares sits, diligently covering her four pretty speckled eggs.

On the reed-fringed banks of the river I have often angled for perch and roach; and began my ornithological pursuits with an old ship's pistol taken from some prize by my father. (This ship's pistol recalls to memory a remarkable incident in my poor father's naval career. The ship on board which he was serving at the time had captured a prize, the crew of which had been bulkheaded off in the forepart of the ship, when about the midnight hour my father, having gone forward to the sick-bay on a visit to the wounded, was startled by the appearance of one of the prisoners who had cut his way through the bulkhead with a knife, which had been concealed beneath the sleeve of his loose sailor's frock, and which knife I recollect well when a boy, for by disarming him my father saved his ship.) I can even now, at this distance of time, recall vividly to mind the enthusiasm of the moment when with it I shot my *first* bird, which happened to be a poor inoffensive hedge-sparrow, or "accentor," hopping in its silent, quiet way from spray to spray in the hedge-row of what was called by the villagers the "Long Lane," in contradistinction to one called the "Short Lane," running at right angles with it, on the left hedge-side of which I also found my first bird's nest, containing the singularly veined, beautiful eggs of the yellow-hammer (*Emberiza citrinella*). After a time a small fowling-piece superseded the old pistol, with which I ventured on somewhat larger game, and shot a starling; and I believe the first bird on the wing was a partridge, if my memory does not fail me.

In the year 1806 we removed to Yarmouth for my education; and subsequently I accompanied my mother to London, to meet my father, whose ship, the *Dolphin* frigate, was refitting at Deptford, where we remained for a few weeks with him at lodgings near the dockyard gate.

On returning to Yarmouth again, and having completed my education, I entered the medical profession, as the only chance now left me of entering upon a naval life. My father's untimely loss, together with the subsequent peace following so soon upon it, proved an insurmountable barrier to my entering the service in the executive line as a midshipman, which had ever been the whole ambition of my existence. Born as I was, within sound of the surf breaking on the sandy seashore, and the report of the evening sunset-gun—fired by the old *Roebuck*, seventy-four, at that time the guardship in Yarmouth roads—the predilection for a sea life, enhanced as it was by surrounding associations, became ultimately a settled conviction with me.

During my subsequent long residence in Yarmouth, my favourite walks were the beach and the old jetty, looking out for the arrival and departure of the few ships of war on the station, and watching with a lively interest their boats approaching in their passage to and fro, or the ships themselves coming to an anchor, or getting under weigh, with the firing of salutes, and other operations.

One beautiful corvette especially, mounting twenty guns, a very model of symmetry in naval architecture—which our transatlantic cousins, the Americans, know so well how to build, called the *Florida*, taking her name from the coast off which she was captured by us during the war—was a great favourite with me, and never since have I met with her equal, either in beauty and symmetry of hull and spars or unrivalled speed. She was commanded whilst on the North Sea station by a Captain Hawtaine.

On every holiday during my school days in Yarmouth I made excursions to Runham, across the intervening tract of marsh-land, some five miles from my residence to that of my grandfather's, where I was born. Starting in the morning and returning the same evening, usually with my gun or fishing-rod, I was mostly accompanied on my way home by my good and considerate old grandfather—with whom I had always been a great favourite—as far as what was called the "Osier Ground" and the "Pickerel Dyke," bridged over by a plank, about half the distance of my journey.

In 1820 I accompanied my mother and two sisters to the pretty rural village of Belton in Suffolk, situated about the same distance from Yarmouth as my own native village, which may be seen across the marshes bounding the opposite horizon. Here we resided for a year, and then removed to Southtown, the charming suburb of Yarmouth.

In the autumn of the year 1821 I went to London, to attend the hospitals of Guy's and St. Thomas's, as a pupil of the late Sir Astley Cooper, returning home for the summer vacation. In the following autumn I returned to the hospitals, and after obtaining the required certificates for passing my examination at the Royal College of Surgeons, the diploma as a member of the college was awarded me on the 6th of December, 1822.

On my return to Southtown I soon afterwards received an answer to the application for an appointment as an Assistant-Surgeon in the Navy, which I had addressed to the Admiralty immediately I became a member of the college. The letter directed me to appear at Somerset House to pass my examination, and on Friday, the 18th of April, 1823, I was passed by Dr. Weir. On the following day, Saturday, the 19th, I received my appointment as Supernumerary Assistant-Surgeon of H.M.S. *Queen Charlotte*, of 108 guns, bearing the flag of Sir James Hawkins Whitshed, Admiral of the Red, at Portsmouth.

Having returned to Southtown, I again left at three p.m. on Thursday, the 24th, by the Yarmouth mail-coach, and at seven a.m. on the following morning I arrived at the "Spread Eagle Inn," Gracechurch Street, from which I proceeded at once by the coach to Portsmouth, reaching the "Fountain Inn," in the High Street, at five p.m., where I slept.

On Saturday, the 26th, I joined H.M.S. *Queen Charlotte*, and ten weeks afterwards, whilst on shore at Haslar Hospital on duty, received an appointment as a Supernumerary for the West India Station, and to embark in the next packet from Falmouth.

On Saturday, the 5th of July, I left the *Queen Charlotte* in a packet-boat for Plymouth, but, owing to boisterous weather, had to put into Portland Roads, where I went on shore and slept at the small inn, and walked across the Chesil Beach to Weymouth. On at last reaching Plymouth, I proceeded from thence to Falmouth by coach, where, arriving too late for the packet, I had to remain for above three weeks at "Selly's Green Bank Hotel" for the sailing of the next packet, during which time I took a trip to Plymouth and back in the *Frolic* sloop-of-war packet.

On Monday, the 11th of August, at two p.m., I bid adieu to England, and sailed from Falmouth in the *Sandwich* packet, commanded by Adoniah Schuyler, with six other passengers, four of them Supernumerary Assistant-Surgeons, who, like myself, were destined to fill up vacancies occasioned by the yellow fever on the Jamaica Station. The two others were Jamaica merchants.

The Diamond Rock, Martinique. (*See page* 192.)

CHAPTER II.

First voyage to the West Indies—Arrival at Barbadoes—Invalided—Return home in H.M.S. *Icarus.*

SAILING from Falmouth on the 11th of August, 1823, we early in the morning of the 25th saw the island of Madeira appearing like a faint blue cloud bounding the horizon. On the 30th we crossed the Tropic of Cancer.

On Friday, the 12th of September, at six p.m., we anchored in Carlisle Bay, Barbadoes, after a fine passage of thirty-two days from England. Early on the following morning I landed for the first time in my life on a tropical shore, walked round Bridgetown, and through several of the plantations in the vicinity. At "Hannah Lewis's Hotel," commanding a delightful prospect of the bay, I rested for awhile, and feasted on some of the delicious fruits of the tropics, washed down by a draught of that refreshing and cooling beverage of the West-

India Islands, "sangree," and, after returning on board, we got under weigh again at five p.m.

Sunday, 14th.—Hove-to off Kingston Bay, in the island of St. Vincent, to send a boat on shore with the mail.

Monday, 15th.—Anchored off George Town, Grenada, and as, after delivering the mail, we sailed again the same evening, I had only just time to run on shore and make a hasty survey of the town.

Monday, 22nd.—We anchored at eleven a.m. in Port Royal Harbour, Jamaica, having been nine days running down the trade-winds from Barbadoes, and exactly six weeks from England. As soon as the packet came to an anchor I repaired on board of the *Gloucester*, seventy-four, bearing the broad pennant of Commodore Sir E. W. C. R. Owen, Commander-in-Chief on the West-India Station. The following ships of war were at anchor in the harbour: *Hyperion* and *Pyramus*, frigates; *Carnation*, of eighteen guns, *Bustard*, ten, *Grecian* cutter, two schooners, and the *Serapis*, convalescent ship. On the following day I was transferred from the *Gloucester* to the latter ship, where I remained until the Commodore appointed me to a sloop of war of eighteen guns, under orders for the Windward Station. During the seventeen days which I remained in Port Royal Harbour I went on shore several times, and once to Kingston.

Thursday, October 9th.—We sailed at daybreak for the Windward Station, on our way calling at Carthagena, on the Spanish main, and then running over to the island of St. Domingo, calling at St. Nicholas Mole, at the western end of the island, and beating up against the trade-wind on the north side. We called at the old city of Cape Hayti, also at St. Juan, Porto Rico; and after our arrival at Barbadoes, our future head-quarters, we were employed cruising after pirates and slavers, with which the Caribbean sea was then much infested, and convoying merchant-vessels through the passages clear of the islands. During this period of a year and nine months I was enabled to see all the islands of the Windward Group, and visit most of them personally, and along the Spanish main from Carthagena to the Gulf of Paria, Demerara, Para, on the mighty Amazon river, and across the Equator. But owing to want of. space, which I wish to reserve for a full account of my voyages to the Poles, I can only afford to give a brief outline here.

Tuesday, July 12th, 1825.—About noon we returned to Jamaica, anchoring in Port Royal Harbour, and, as my health had suffered severely from an attack of yellow fever, and a very uncomfortable ship, ill-suited for a convalescent, I invalided, and was ordered a passage home in H.M.S. *Icarus*, one of the old ten-gun brigs, called "coffins," from so many of them having foundered at sea, being never heard of after. From their peculiar build and size, if taken aback in a squall, they were liable to go down stern foremost. We were crowded with supernumeraries, and having a prolonged passage, although a fine one, ran short of water. But the Commander, John George Graham, brother of the Scotch Professor of Botany of that name, was an excellent fellow, and did everything in his power to make all of us as comfortable under unavoidable privations as was possible.

Thursday, 28*th*.—At daybreak we sailed from Port Royal, and, after a somewhat long passage, anchored at Spithead at eight a.m. on Sunday, the 11th of September. At four p.m. I landed at Point, and slept at the "Star Tavern," after two years and one month's absence.

Summary.—Of all the beautiful islands forming the Windward Group, Martinique struck me as the gem of the Caribbean Sea. I first saw it on the 30th of September, 1824, at ten a.m.; but having only hove-to to send a boat on shore, in which I had not an opportunity of landing, I made a sketch from the ship's deck. Subsequently, however, on the 24th of the November following, we paid a second and last visit to the island, anchoring off St. Pierre at nine a.m., where we found a French man-of-war brig of sixteen guns at anchor. On the following morning, at early dawn of the 25th, I went on shore, and made an excursion to the fine botanic garden on the right of the town. It is a delightful spot, irrigated by a splendid water-fall descending from the summit of the mountain, at the base of which the tastefully laid-out garden is situated; the loud musical sound of the falling water reaches the ear before the cataract itself comes in view. I had an interview with the venerable old French botanist, who was accompanied by a bevy of beautiful young French girls, his pupils, to whom he introduced me. The streets of St. Pierre have a running stream of water in the centre, rendering them remarkably cool and pleasant. I went to the hotel, where I got some fine ginger and Guava jelly preserves.

At Bermuda I met with an old messmate of my poor father's, James Squire, the master-attendant of the dockyard, upon whom I called. The characteristic feature of vegetation of the islands is the cedar (*Juniperus Barbadensis*). Among the birds of these cedar groves the most striking are the "Blue Bird" (*Sialia sialis*), the "Red Bird" or "Summer Bird" (*Pyranga æstiva*), and the "Red Cardinal" (*Cardinalis virginiansis*). At sea the "Sheerwater" (*Rynchops nigra*) is a great wanderer, as he glides over the crests of the rolling waves, with his long pointed wings uplifted at an angle with his body, and the black "Man-of-war" bird, with his bright crimson crop (*Pelicanus frigata*), soaring on high with great extent of wing; and further south may be seen the pretty "Blue Petrel" (*Procellaria cærulea*) and many other sea birds.

CHAPTER III.

On the home station—Employed in the coast blockade—Apply to join Captain Parry—Appointed to the *Hecla*—Pass examination at Royal College of Surgeons, and at Naval Medical Board—Depart for Spitzbergen.

On Sunday, the 18th of September, 1825, I left the "Star" and proceeded by the Portsmouth coach to London, where I arrived the same evening.

Friday, 23rd.—I passed the customary examination at the Admiralty, as an invalid, and on Wednesday, the 28th, I received an appointment to H.M.S. *Ramillies* of seventy-four guns, stationed as the coast blockade ship in the Downs; and on the following day at eight a.m. I proceeded by coach to join her, arriving at Deal at seven p.m.

Friday, 30th.—Reported myself at the Admiralty Office, as it was blowing too hard for any boat to get off to the ship.

Having on Saturday, the 1st of October, joined the *Ramillies,* on the 5th I volunteered for her tender, the *Antelope* cutter of 120 tons, to cruise in the North Sea during the winter months, encountering heavy gales of wind and a boisterous sea off the coast of Holland, and saw the Texel lights, expecting every moment to be driven on shore during one dark, tempestuous night. We were driven over the Newark Sands, when a Winterton boat coming off to us, I embraced the opportunity thus afforded to land in her, and send off a pilot from Yarmouth. Those splendid boatmen succeeded in a marvellous manner to beach their boat through a tremendous surf and heavy rollers. I managed, after some difficulty, to get a conveyance from Winterton to Yarmouth in a smuggler's cart, reaching Southtown between the hours of eight and nine p.m. on Wednesday, the 9th of November; and, having sent off a pilot to the cutter, I remained for a brief visit with my mother and sisters.

Friday, November 18*th.*—At nine a.m. I went by coach to Norwich, and slept at the Norfolk Hotel. On the following morning I took the coach, passing through East Dereham, Lynn, Wisbeach, &c. On Sunday, the 20th, at six p.m., I crossed the Humber in the steamboat, and rejoined the *Antelope*, lying in the dock-basin, Hull, on the same evening.

Sunday, December 18*th, at eleven a.m.*—After a month's very agreeable sojourn at Hull, and making some rambles in the vicinity, we got under weigh again for the Downs, where we came to an anchor on Saturday, 24th, at two p.m.

Sunday, June 11*th,* 1826.—Having received orders to relieve the medical officer in charge of the Epple Bay Station, who, from illness, was incapacitated from performing his duty at the shore stations along the coast, I left in the *Ramillies* galley, and landing at Margate Pier at seven p.m., I walked to the Epple Bay watch-house, where I was kindly welcomed by the officer in charge, Lieutenant John Stephen, and his wife and sister, Miss Sarah Warner. Instead of having to seek lodgings for a temporary stay, they insisted on my remaining under the same roof with them and joining their mess, having a bed made up for me on the couch, as there was not a spare bedroom. Thus I passed the summer very agreeably, visiting the various stations under my charge along the cliffs on either side, from Ramsgate to the St. Nicholas Station, embracing Westgate, Newgate, Kingsgate, Broadstairs, and Margate Stations; when, having received orders to return on board, on Friday, the 25th of August, at nine a.m. I embarked in the *Julia* tender for the Downs, and at four p.m. rejoined the *Ramillies.*

Monday, September 4th.—I wrote a letter to the Admiralty requesting to be superseded, with the intention of making every effort in my power to get out with Captain Parry in his proposed expedition to attempt to reach the North Pole.

Wednesday, 13th.—I was superseded, and on Saturday, the 16th, at five p.m., I left the *Ramillies,* after about a year's service on the coast blockade.

Monday, 25th.—I made personal application at Somerset House to be appointed to the North Polar Expedition, when I was told by Dr. Weir, the medical commissioner, that Captain Parry had the selection of his officers. I therefore on the following day obtained an interview with Captain Parry himself at the Admiralty. He said, having already many volunteers on his list, and one especially recommended to him for the appointment by the captain-superintendent of Haslar Hospital, he could only enter my name as a candidate. Undismayed, however, at not getting a decisive answer, I sought to obtain the influence of my old tutor, Sir Astley Cooper, to back me up; and on Friday, the 29th, I had an interview with him at Spring Gardens. Sir Astley received me most kindly, expressed an interest in my enterprise, and said he thought he could manage this for me with Parry, whose brother, Dr. Charles Parry, he told me, had been a pupil of his. So taking up a pen and sheet of paper, he wrote a line or two to Captain Parry, handing it to me open and unsealed, which, on the following day, Saturday, the 30th, I placed in Captain Parry's hands at the Admiralty, when he said, "This speaks volumes for you, and as you appear to be so enthusiastic in the service, I shall nominate you at once; therefore lose no time in going into Norfolk to take leave of your friends." Sir Astley and myself were both natives of the same county; Sir Astley's father had been rector of Great Yarmouth, where he himself was born.

Wednesday, October 11th.—I again called on Captain Parry at the Admiralty, when I was told by him that the *Hecla* would be commissioned on the 14th of November.

Thursday, 12th.—At Captain Parry's request I had an interview with Mr. Charles James Beverley, his former medical officer in the late expeditions in search of the North-West Passage, and who was going out in the forthcoming expedition, nominally as naturalist, to enable him to get reinstated in his rank on the Navy List, from which his name had been removed for refusing to serve when called upon. I saw both himself and his wife at his residence in the North Crescent, Bedford Square.

Monday, 16th.—On finding myself appointed to the *Hecla* by Captain Parry, I called on Sir Astley Cooper, to acquaint him with it, and to thank him.

Saturday, 21st.—At five a.m. I left the "Spread Eagle," Gracechurch Street, by coach, for Yarmouth, and reached Southtown, the residence of my mother and sisters, about ten p.m.

Monday, November 13th.—At three p.m. I left Southtown by the mail coach, and arrived in London on the following morning at eight o'clock; and on Wednesday, the 15th, took up my appointment to H.M.S. *Hecla,* joining her about noon on Saturday, the 18th, when her pennant was hoisted by Mrs. Parry. I paid a visit to my old cutter, the *Antelope,* lying off the dockyard.

Wednesday, 22nd.—This afternoon got my baggage on board the *Hecla's* hulk, *Heroine.* Slept at an inn in Deptford; and on the following day took up my quarters on board the hulk.

Wednesday, December 20th.—Received the *Hecla's* supply of medicines and stores from the Victualling Yard at Deptford.

Tuesday, January 16th, 1827.—I went to town and

ordered a double-barrelled gun at Wilkinson's, of Ludgate Hill, to be made expressly for me, with case and everything complete.

Monday, 22nd.—Captain and Mrs. Parry came on board the *Hecla*, and remained until Thursday, the 25th.

Saturday, February 10th.—I went to Wilkinson's and saw the barrels of my new gun, and selected a handsome walnut stock for it.

Tuesday, 20th.—The river was full of floating ice, and partially frozen over near its banks. On the following day I went to town, and got a letter from the naval Medical Commissioners to the College of Surgeons, applying for an examination, for notwithstanding I was already a member of the College, the regulations of the service required at the expiration of three years a certificate from the College of fitness for the rank of full surgeon in the navy. On returning to Deptford in the evening, I found the lower water-gate so beset with ice as to prevent the wherries from getting off, and I was compelled to go round to the dockyard and get on board the hulk by means of a kind of chair, made for the purpose, and drawn along a rope fixed between the dockyard and hulk, and thence from the hulk in a boat to the *Hecla*.

Friday, 23rd, at six p.m.—I passed the required examination at the Royal College of Surgeons, and went to Covent Garden Theatre afterwards. On the following day I obtained the certificate from the College, and took it to Somerset House. On Monday, the 26th, at one p.m., I was again examined by Dr. Weir, the Medical Commissioner, for the rank of full surgeon, and received the certificate at the secretary's office on the following morning.

Tuesday, March 6th.—Went to the Victualling Yard at Deptford, and ordered the remainder of extra supply of medical stores for this particular service, which came on board on Thursday, the 8th.

Monday, 12th.—I went to town, and tried my new gun at Wilkinson's shooting-ground, with Mr. Wilkinson, jun., and my young friend and brother officer, Pierce Power, R.N., also the son of an old naval surgeon, and the friend of the celebrated Barry O'Meara, surgeon of the ship which took the great Napoleon to St. Helena, where he became his medical attendant, and published an account of his exile in the "Voice from St. Helena." Poor Power sailed for the West Indies in the *Harpy*, ten-gun brig, which was never afterwards heard of, and doubtless foundered at sea.

Saturday, 17th.—The ship's company had a dinner given them on board by Captain Parry; and a party of the captain's friends, ladies and gentlemen, came on board and witnessed the proceedings.

Wednesday, 21st.—We had a number of visitors on board to-day to see the ship; and on the following day the numbers were greatly increased.

Friday, 23rd.—The ship's company mustered at divisions, with the officers in cocked hats and swords, to receive the Lords of the Admiralty, with a great number of visitors to see the ship. On the day following, the ship's company were shifted from the hulk to the *Hecla*, the boats were got on board, and everything made ready for sea. Our decks to-day were so crowded with visitors, both ladies and gentlemen of all ranks, that it became extremely difficult to make one's way along either upper or lower deck, or from one cabin to another.

On Sunday, March 25th, at ten a.m., we cast off from the hulk, and went down the river.

CHAPTER IV.

Return—Promoted—Study botany—A year's leave—Appointed to the *Hyacinth.*

Thursday, November 1st, 1827.—The *Hecla,* on her return from the Arctic seas, having been paid off at Deptford, we had a parting dinner at Hill's "Ship Tavern," Charing Cross, and I took up my quarters at my old lodgings, No. 4, Duke Street, Adelphi, and applied for two months' leave of absence to prepare my accounts for office.

Tuesday, 20th.—I applied for my promotion, by forwarding a memorial of my services to the Lord High Admiral, his Royal Highness the Duke of Clarence, through whom I had the gratification to receive it, promptly and at once, without having my memorial backed up by any interest or influence whatever, either from my old captain or the Medical Board; my commission, which was dated the 27th of November, appointing me to H.M.S. *Nelson* for rank.

Friday, December 11th.—I applied for six months' further leave on half-pay, and obtained it.

Saturday, 12th.—I left town by the seven p.m. Norwich coach for Yarmouth; reached Norwich on the following morning, and the same evening took the coach to Yarmouth, where I arrived late, and walked home to Southtown.

Monday, March 10th, 1828.—I left Southtown by the

five a.m. coach for London, arriving at the "Spread Eagle," Gracechurch Street, at nine p.m., and from thence went to my lodgings, No. 4, Duke Street, Adelphi.

Thursday, 20th.—I purchased Captain Parry's quarto narrative of our voyage at Murray's, in Albemarle Street.

Tuesday, April 1st.—I took the coach to Kew Gardens, and walked from thence to Richmond and Twickenham, returning by the Twickenham coach to town. On the following day I took the coach to Camberwell and Denmark Hill, and walked from thence through Dulwich back to town. My object was to find a suitable place of residence for my mother and sisters.

Thursday, 3rd.—I went by coach to Wimbledon, alighted on the common, strolled through the park to Wandsworth and Clapham, returning to town by the Clapham coach. On the following day I took coach to Hampstead, and walked back to town.

Monday, 7th.—I received from my agents, Messrs. Cook and Halford, two proof copies of the battle of Navarino, by my old friend, Sir J. Theophilus Lee, to which I had been a subscriber.

Wednesday, 9th.—At ten a.m. I left town by the Hastings coach from the "Bolt-in-Tun," Fleet Street, for an excursion along the South Coast. Arrived at the "Castle Hotel," Hastings, at seven p.m., and on the following day I rambled in the vicinity, along the beach, to where the beautiful town of St. Leonard's now stands. The stone called the Conqueror's Table had only just been raised from its ancient site, on which two hotels are now, with a row of lodging-houses, building. The next morning I took the coach to Brighton, took a run on the Chain Pier, and slept at the "Pavilion Hotel." On the 12th, at eleven a.m., I took the coach to Portsmouth, reaching the "Fountain," my old inn, at six p.m., where I slept. On the following day I crossed over to Ryde by the steamer, passing close to my old ship the *Hecla*, at anchor off Spithead, and now bound for the coast of Africa, after her buffetings with the ice in the Arctic seas. From Ryde I went on to Newport and Cowes.

Tuesday, 15*th*.—I took the sailing-boat to Southampton, where I arrived at 5.30 p.m. and slept at the "Sun Tavern." On Thursday, 17th, I left Southampton by the Union coach at 8.30 a.m. for London, where I arrived at five p.m.

Tuesday, 29*th*.—I called on Captain Parry at the Admiralty, where he was now the hydrographer, and had my journals and sketches of the late Polar voyage returned to me, from the Record Office, and I afterwards entered my name at Coutts' bank, as a subscriber to the Clarence medal.

Monday, *May* 5*th*.—I took the coach to Dorking, through Kennington, Clapham, Tooting, Merton, Ewell, Epsom, Leatherhead, &c., returning to town at seven p.m. On Wednesday, 7th, I took the coach to Clapham and back, having walked through Merton.

Friday evening, 9*th*.—I engaged a place in the coach for Yarmouth, slept at the "Spread Eagle," Gracechurch Street, and at five a.m. on the following morning I started for Southtown, where I arrived at ten p.m.

Tuesday, 27*th*.—I removed my mother and sisters into a newly-built, semi-detached villa, Southtown Terrace, on the London Road; the river running alongside of the road, with a fine open prospect of the South Denes, the sea in the distance in front, and a wide expanse of meadow-land and marsh in the rear. Having been unsuccessful in my search around the suburbs of London for a suitable home for them.

Thursday, *June* 12*th*.—I returned to town by the five a.m. coach, which reached the "Spread Eagle" in Gracechurch Street at nine p.m., and I proceeded to my old lodgings in Duke Street, Adelphi.

Monday, 15th.—I applied for a renewal of leave, my state of health having been in a far from satisfactory state, and on the 20th I obtained an extension of three months. Next day I visited the Exhibition of Paintings at the Royal Academy at Somerset House, and at seven p.m. left town by the Telegraph coach from the "White Horse," Fetter Lane, arriving at Southtown on the following morning at eleven o'clock.

Tuesday, July 22nd.—The twenty-eighth anniversary of my birthday, even to the day of the week, as I was born on a Tuesday. I set some of the Spitzbergen seeds in the garden after a shower of rain, but only three or four plants of one kind sprang up, and these merely to wither away and die. To-day I read Captain Parry's narrative through for the first time.

Monday, September 15th.—I wrote a letter to the Medical Board, reporting the state of my health, and on the following Thursday received an answer, granting me a year's further extension for the recovery of my health, and I remained with my mother and sisters in Southtown Terrace for the winter.

Tuesday, May 5th, 1829.—I took my departure from Southtown by the coach for London, arriving at nine p.m., and slept at the "Salopian Hotel," Charing Cross. On the following day I entered my name at the London University to Dr. Lindley's new course of lectures on Botany, on the Natural System, and at eleven a.m. attended the first lecture.

Tuesday, 19th.—Purchased Professor Lindley's Introductory Lecture, and in the afternoon went to Woolwich and paid a visit to my old shipmate, James Ross, on board his little steamer the *Victory*, bound for the Arctic regions, and he introduced me to his uncle, old Sir John, on the quarter-deck.

Wednesday, June 3rd.—I met Sir Edward Parry in Regent Street, who told me that he was going out to

Australia as Commissioner to the Australian Agricultural Company.

Wednesday, 24th.—I visited the Exhibition of the Royal Academy at Somerset House, and on Monday, the 29th, gave in my name to Professor Lindley, at the Botanical Theatre, for a note of introduction to the Chelsea Botanic Garden.

Thursday, July 2nd.—I obtained a ticket of admission from Apothecary's Hall to the garden. On Saturday, the 11th, I breakfasted with Professor Lindley at his residence on Turnham Green, and accompanied him on a visit to the Horticultural Society's Gardens. On Friday, the 17th, I paid a visit to the Botanic Garden at Chelsea.

Wednesday, August 5th.—The course of lectures on botany having concluded to-day at the London University, on the following day I made application to Dr. Weir for an appointment on the Coast Blockade, and on Saturday, the 8th, entered my name at Maude's the Navy Agents, as a subscriber to the bust of the late Lord Melville, who had done much for our department.

Friday, September 4th.—I received an appointment to the Mediterranean station, a station above all others I was most desirous of serving upon; but this appointment being only as a supernumerary, and a temporary one, I got it cancelled.

Wednesday, 9th.—I went to Yonge's, the watch-makers, in the Strand, and found that the silver hunting-watch I had previously ordered to be expressly made for me was finished.

Thursday, October 8th.—I went to the Adelphi Theatre and saw Cooper's "Red Rover," "The May Queen," and "York and Lancaster."

Saturday, 10th.—Learnt at the Admiralty that the *Hyacinth*, a new corvette, of twenty guns, lately launched at Plymouth, was commissioned, and I applied for her,

I was appointed to her on the 12th, and took up my commission on the following day.

Tuesday, 20th.—At 7.30 a.m. I left Charing Cross by the "Hero" coach for Portsmouth, and at 6.30 arrived, and slept at the "Blue Posts Inn" for the night.

Wednesday, 21st.—I went on board the *Ganges* hulk, and on the following day on board the *Rapid*, and at seven p.m. left Portsmouth by the *Sophia Jane* steamer for Plymouth, arriving on the following day at eleven a.m. Landed at Mutton Cove, and took up my quarters at the "London Inn." Found my old cutter the *Antelope* lying in Hamoaze. In the afternoon I walked out to the Citadel to see an old acquaintance of mine, Dr. Forrest, of the 23rd Fusileers.

Saturday, 24th.—Reported myself at the Clerk of the Cheque's Office in the dockyard, and went on board the *Hyacinth*, lying off the dockyard.

Sunday, 25th.—I dined at the mess of the 96th Regiment at the citadel.

Monday, 26th.—Reported myself at the Admiralty Office, and on the following day the *Hyacinth's* pennant being hoisted, I joined her, and the next day went on board her hulk the *Diadem.*

Monday, November 2nd.—Captain Robert Milborne Jackson having joined, I learnt from him, to my great disappointment and mortification, that we were destined for the West India station instead of the Mediterranean, as all had expected.

Friday 6th.—I took up my quarters on board the hulk *Diadem*, having been living on shore ever since my arrival in Plymouth.

Monday, 16th.—All the members of our mess dined on board together for the first time.

Monday, 23rd.—I called at the Admiral's office, and received an answer to my application for leave of absence, granting me three weeks. On the following day at ten a.m. I left "Weakley's Hotel," Devonport, by the "Defiance" coach, for town, arriving there between three and four p.m. on the next day. Went to my old lodgings in Duke Street.

Thursday, 26th, and Saturday, 28th.—Dined at my old place the "Woodstock," and at seven p.m. of the latter day I left by the "Telegraph" coach for Yarmouth, reaching Southtown at eleven a.m. on the following day.

Tuesday, December 8th, at 3.30 p.m.—I took leave of my friends at Southtown, starting by the London mail. I arrived in London at 7.30 a.m. on the day following, proceeding to Duke Street, Adelphi.

Tuesday, 15th.—Had a seal with my initials engraved on it made out of the Spitzbergen "Rose-Quartz."

At 3.30 p.m. of the 17th I left the Western Office in Regent Street by the Plymouth coach, and reached the "London Inn" on the following day at nine p.m.

Saturday, 19th.—Rejoined the *Hyacinth.*

Thursday, 24th.—Went to the Naval Hospital for the supply of medicines, and passed the 25th, Christmas Day, on board.

Wednesday, January 1st, 1830, at eight a.m.—The officers and ship's company were turned over from the hulk to the *Hyacinth.*

Monday, 11th.—The crew were paid their wages. The captain's wife, with her two pretty young daughters, came on board to take a final farewell of the ship, and on their return in the yacht I accompanied them on shore.

Nassau Light-house, and *Hyacinth's* Boats in chase of a Slaver.
(*See page* 209.)

CHAPTER V.

To the West Indies again—Chase of a Slaver off Nassau—Again invalided—Return in the *Arachne* home.

Tuesday, 12*th of January*, 1830.—We got under weigh from Plymouth Sound, going out by the west channel with a strong breeze from the N.E. The night was fine and moonlight; saw the "Eddystone Light" and, during the first watch, the "Lizard." We had a very rough and boisterous passage out, a succession of heavy westerly gales, and in passing between the Great Canary Island and that of Teneriffe, on the 27th, we encountered a tremendous squall—one fearful gust of wind nearly threw the ship on her beam ends—and a heavy green sea she shipped at the time swept her decks like a deluge fore and aft. Lee guns and gunwale were all under water, and her yard-arms to leeward for a few moments, which seemed like an age before she righted, rested on the vortex of the turbulent waters which raged around and threatened every instant to engulf her. The three top-sails were on the caps at the time the squall struck her. About midnight the night was dark, and heavy showers of rain, accompanied by incessant lightning, every flash of which exhibited the high bold land on either side in strong relief, under a canopy of dense black clouds, so that we were thus prevented from calling at Teneriffe as was intended.

Monday, February 1st.—We crossed the tropic, when the usual ceremony was gone through, by old Father Neptune and his crew shaving and ducking all the greenhorns who had not crossed before; but has been too often described to be repeated here.

February 10th.—We anchored in Carlisle Bay, Barbadoes. Since crossing the tropic we have been accompanied by shoals of flying fish. During the week of our stay here I went on shore several times, dined at Betsey Austin's hotel, and had "sangree," that agreeable, cooling draught of the tropics, at Betsey Lemon's, both celebrated characters in Bridgetown, their taverns having been for years the resort of both the navy and army.

On the 17th we sailed for Trinidad, passing through the Bocas's by the Umbrella passage, and on the 19th in the afternoon came to an anchor off Port d'Espagne. Here we found Admiral the Hon. Elphinstone Fleeming, in the *Barham* frigate of fifty guns, and as soon as we had secured the ship, I accompanied Captain Jackson in his gig on shore, and we had a pull for four or five miles through such torrents of rain as are only to be met with in the tropics. On landing, went to the hotel in the English part of the town.

Sunday, 21st, at five p.m.—We weighed anchor in company with the *Barham*, under a salute of seventeen guns from the shore, which was returned by the *Barham*. We repassed the Bocas's with a fresh breeze, and outside we parted company with the admiral to proceed to Jamaica, calling at Grenada and St. Kitt's on our way. I went on shore at both places with the captain, and accompanied him in his visit to the governor of St. Kitt's.

March 4th.—Beating up along the lower Whitehorses

for Port Royal Harbour, anchoring for the night off the Palisades, and on the following morning came to an anchor in the harbour, when I accompanied Captain Jackson in his gig on shore. Went to the well-known "John-a-Ferong's" tavern and stores, which has long been the favourite resort of naval officers; and I went afterwards to the Naval Hospital.

Sunday, 7th.—Went in the gig to Kingston at five a.m., and managed to get on shore in rounding Palisade Point, but got afloat again immediately; reached Kingston in about an hour, landing at the market-place, which I found abundantly supplied with provisions and fruits of all kinds, and I returned on board to breakfast.

Tuesday, 9th, at two p.m.—Sailed for New Providence to refit; and on Thursday, the 18th, we crossed the Tropic of Cancer, passing within about thirty miles of the island of St. Salvador, the land first sighted in Columbus's first voyage.

Saturday, 20th.—Anchored off the town of Nassau, New Providence, and at seven a.m. on the following day I landed for the first time at the Bahama Islands, and strolled from the market-place in the west to the Eastern Church in the east end. Went to Mrs. Baldwin's establishment to see the beautiful shell-work for which Nassau is so celebrated, all made by her nieces and their pupil, a pretty sylph-like young Creole girl, Sarah Elizabeth Robinson, a native of the island, here called a "Conch," after, I suppose, the shell of that name so abundant in the Bahamas. This young girl's work in flower wreaths, baskets, and bracelets was all done in the most tasteful style, several specimens of which I brought home with me.

Tuesday, 23rd.—This morning, after going on shore to order some shell-work at the Baldwins, returned on board at six p.m. to dine with Captain Jackson. In the midst of dinner a slaver was reported within sight of the island, and chased by the *Kangaroo* schooner. Our cutter and pinnace were instantly manned to go in pursuit of her, and to endeavour to cut her off in the direction she was steering. Having volunteered my services on the occasion, and being accepted by the captain, I accompanied Sturt, our first lieutenant, in the cutter, and in the hurry-skurry of the moment, after shoving off, as we made sail, I discovered that the ammunition, with the ship's pistols and cutlasses, had been thrown into the bottom of the boat in hapless confusion, which, whilst my companions in this exciting chase were attending to the sails, I arranged and distributed the arms and ammunition; and after standing out to sea for some twelve or fourteen miles, and the darkness of night shrouding all around, left us no other alternative than to bear up, and shape our course for the island by the lighthouse on the point, getting on board again after midnight, without falling in with the prize we had hoped to have captured and brought back with us. We subsequently learnt that she was full of slaves, mounting fourteen guns, and well manned; so that had we succeeded in boarding her, it would doubtless have been smart work for our two boats with such few hands to have carried her.

Thursday, April 1st.—I dined at Dunmore House with the officers of the 2nd West India Regiment at their mess.

On Monday, the 12th, the ship's company were quartered on shore at Hog Island, whilst the ship was being painted; and with the rest of the officers I took up my quarters at Mrs. Baldwin's boarding-house. And next day again dined at Dunmore House with the officers' mess.

Friday, 30th.—I attended an invaliding survey on board the *Mersey;* and my own state of health having been anything but satisfactory since my return to a station where I had formerly suffered so much from yellow fever and the climate as to leave me no other alternative than to invalid home, I at once requested the officers who with myself were holding the present survey, to include me in it, and I was consequently invalided the same day, and was ordered a passage home in the *Barham*.

On Saturday, May 1st, I left the *Hyacinth*, and on the 12th she sailed for Bermuda, and the *Arachne* of eighteen guns having arrived in the interim, I was on the 17th ordered a passage home in her instead of in the *Barham*. I went to the public ball in the evening, to which I had been several times during my sojourn in Nassau, a very pleasant place and abounding in hospitality.

Thursday, 20th.—This forenoon I took my baggage on board the *Arachne*, and passed my last evening at Nassau at a friend's house at the east end of the town, a picturesque spot, over which the graceful cocoa-nut palms waved laden with fruit in clusters, suspended beneath their feather-like tufts of large leaves which crowned the tall erect stems.

On the following day, the 21st, about eleven a.m., I took my final leave, and with some lingering regret of a pleasant place, in which I had passed many happy hours, and may never again visit, I started for "Cochrane's Anchorage" in Mrs. Baldwin's boat, and was accompanied by a brother officer, who had also invalided, and was to go home in the same ship as myself. At four p.m. we reached the *Barham*, and the *Arachne* not having as yet reached the anchorage, I remained the night on board the *Barham*. The next morning at eight a.m., on the arrival of the *Arachne*, I at once joined her. The *Blossom* was the only ship of war left at Nassau.

Saturday, 22nd.—Weighed in company with the *Barham*, anchoring again at night; weighing again next day, and coming to an anchor at night.

Monday, 24th.—A fine fresh breeze springing up,

carried us through the "Six Shilling" channel, and well to windward of the "Hole-in-the-Wall," and saw the beacon on the island of Eleuthera.

Thursday, June 3rd.—In latitude 37° N. In the evening, whilst going ten knots with a strong breeze, a flying-fish of unusual size, eighteen inches in length, wings ten inches in length and six inches in breadth, and body six inches in circumference, came on board over the weather quarter, which I preserved in spirits.

Tuesday, 8th.—Another flying-fish, but much smaller than the last, fell at my feet, whilst walking the quarter-deck this evening. Stormy petrels have been following in the wake of the ship ever since we left Nassau.

June 9th and 10th.—Blowing a heavy gale of wind, under close-reefed top-sails and fore-sail, with the hatches battened down.

Saturday, 12th.—Passed close to the island of Flores, its high and barren-looking summits capped with clouds, with numerous waterfalls from the ravines and down the mountain sides. With Flores still in sight we passed Corvo, a small, barren-looking island, having a dark, steep, rugged shore, deeply excavated with caves.

Wednesday, 23rd, about four p.m.—The Scilly Islands appeared to leeward on the port-bow, very low, with the lighthouse on one of them. At seven p.m. we made the Land's End, through thick and hazy weather. Saw both the West and South Land's End, and the Longships' rock. Fired a gun to warn the *Barham* of breakers ahead, on the Wolf rock. The wind having suddenly changed against us, we had a beat up channel, blowing fresh, with thick, hazy weather, rain, and lightning, the land appearing enveloped in dark clouds and mist.

Saturday, 26th.—We passed the Needles about noon, afterwards Yarmouth and Cowes, anchoring off Spithead at two p.m., after a passage of five weeks, during which time we never once lost sight of the *Barham*. Admiral

Fleeming fired a salute going into the anchorage. At four p.m. I landed at the Custom-house, Portsmouth, and took up my old quarters at the "Star Inn" Point.

Sunday, 27th.—Went on board the *Arachne*, and on my return to the shore reported myself at the admiral's office. Last evening the news of the king's death arrived.

Monday, 28th.—At 10.30 a.m. I left the "Star" by the "Rocket" coach for London, arriving at Charing Cross at seven p.m., where I slept at the "Ship Tavern."

Friday, July 2nd.—Made my appearance with several other officers as invalids at the Admiralty. After which we had a parting dinner at the "Ship Tavern," Charing Cross.

The Church, East end of Nassau.

CHAPTER VI.

A year on half-pay—Attend Professor Jameson's Course of Lectures on Natural History at the Edinburgh University.

Saturday, July 3rd, 1830.—I went to the Exhibition of the Royal Academy at Somerset House, and again on the 19th; and on the following day, at seven p.m., I left the "White Horse," Fetter Lane, by the "Telegraph" coach for home, after getting my name noted at the Board for a coastguard appointment. At eleven a.m. next day I arrived at Southtown Terrace.

Monday, November 1st.—At five a.m. I left Southtown Terrace by the "Star" coach for London, and slept at the "Ship Tavern," Charing Cross; and on the following day I took up my quarters at 15, Duke Street, Adelphi.

Thursday, 11th.—Had my name noted for the Mediterranean or South American stations, and having on the following evening received a note acquainting me with two vacancies on the coast blockade, on asking for one of them I found they had both been filled up. Only Scotchmen have any chance with the head of our department.

Sunday, 14th.—Finding I had no prospect for the present of any suitable appointment, and that Professor Robert Jameson's course of lectures on Natural History at the Edinburgh University was about commencing, I made up my mind to pass the winter in the modern Athens, and to attend them. I consequently, at two p.m., embarked from Miller's Wharf, on board the *Earl of*

Hopetoun packet, for Leith; on the 17th anchored in Leith roads; and on the following morning I landed, and walked to Edinburgh by the Leith Walk. This being my first visit to Scotland. I breakfasted at the "Crown Inn," Princes Street, and afterwards commenced a search for lodgings, which I found at No. 3, Montagu Street, on the southern outskirts of the new town, and they turned out very comfortable.

Friday, 19th.—I went to the University and entered my name to Jameson's winter course of lectures, attending the first one at two p.m. On the following day I paid a visit to the Museum of Natural History.

Monday, 22nd.—I was introduced to the George Street Subscription Reading Rooms, for a fortnight, and at the expiration of that time I became a subscriber for the next six months.

Tuesday, 30th.—I entered my name at the library of the University for a matriculation ticket.

Thursday, December 2nd.—I commenced attending Dr. Mackintosh's courses of lectures on the Practice of Medicine and of Midwifery, and also Dr. Lizar's course of lectures on Anatomy, for both of which, as a medical officer in the navy, I had tickets presented to me free.

Friday, 3rd.—I attended Dr. Campbell's lecture on the Practice of Physic. And on Monday, the 6th, I attended Dr. Fletcher's lecture at the Institute of Medicine. On Saturday, 25th, Christmas Day, between seven and eight o'clock in the evening, I witnessed a display of the aurora, or the Northern Lights. On Monday, the 27th, I made an excursion to Duddington Lock, which was frozen over, and a vast number of people skating on it.

Tuesday, February 1st, 1831.—The frost set in excessively severe, accompanied by a heavy snow-drift, and keen easterly wind.

Monday, 21st.—I attended Dr. Mackintosh's lecture on Medical Jurisprudence, and on March 4th, Friday, I attended his last lecture on Medicine and Midwifery, and also Dr. Lizar's lecture on Anatomy.

Monday, 7th.—After visiting the Museum of Natural History, in the evening I saw a most brilliant display of the aurora borealis, shooting from east to west. On Thursday, the 24th, I made an excursion on foot to Bonnington and Newhaven.

Saturday, April 2nd.—Visited Professor Jameson's own collection of Natural History, in the rooms opposite to the side on which the public museum stands; and at two p.m. accompanied him and his class on an excursion to the Salisbury Crags—the geological structure and mode of formation of which, with the hills adjacent, he explained on the spot, and we returned at 3.30 p.m.

Sunday, 3rd.—I attended forenoon service at Hope Park Chapel, and the afternoon service at St. John's Chapel. April 6th, on an invitation from Professor Jameson, I attended a musical soirée, at his house in the Royal Circus, in which his niece, a clever, interesting young girl, just entering her teens, played on the harp, and has a talent for composing.

Saturday, 9th.—I went with Professor Jameson and his class, on a geological excursion to the summit of Arthur's Seat; started at two, and returned at four p.m.

Wednesday, 20th.—Visited the museum and completed the list I had been making during the winter, of the specimens of natural history contained in it.

Sunday, 24th.—Attended the afternoon service at St. Mary's Church, and on Monday, the 25th, paid my last visit to the Natural History Museum.

Tuesday, 26th.—Made a pedestrian excursion to Leith, through Warriston, Porto-Bello, Fisherrow, and Musselburgh, returning by Duddington.

Wednesday, 27th.—Ascended the Pentland Hills, passing through the village of Morningside, and returning by the village of Collington and the "Hunter's Tryst."

Sunday, May 1st.—Attended morning service at St. George's Church.

Monday, 2nd.—At eight a.m. I went to the Botanic Garden, and attended Professor Graham's introductory lecture on Botany. The Professor is a brother of Captain John George Graham, R.N., with whom I returned in the *Icarus*, ten-gun brig, from the West India station in 1825. At one p.m. I attended Dr. Hope's introductory lecture on Chemistry; and at two p.m. Professor Jameson's introductory lecture to the summer course of Natural History

Tuesday, 3rd.—I paid a visit to the old Castle, and to Holyrood House; saw the picture-gallery and the apartments of Mary Queen of Scots. On Wednesday, the 4th, I went to Leith, and took my place in the *Duke of Buccleuch* smack for London.

Saturday, 7th.—About seven a.m., being a fine morning, I walked to Leith by the Leith road, or "walk," as it is called. And at ten a.m. the packet started, passing in sight of most of the places along the coast; and on Tuesday, the 10th, about six p.m., passed through Yarmouth Roads; and I had a fine view of the old jetty, with Nelson's monument, and the houses on the Southtown road. Arrived at London on the following day at one p.m.; landed at the wharf at three p.m., and took up my quarters at my old lodgings, 15, Duke Street, Adelphi.

Monday, 16th.—I wrote a letter to the Admiralty, to be employed. On the 30th I called on Mr. Barrow, the Secretary to the Admiralty, and on the Hydrographer, Captain Beaufort, at his office.

Tuesday, June 21st.—I was offered an appointment at Riga, which I declined.

Having now fairly taken up the pursuit of natural

history, in addition to my ordinary professional duties, and prepared and qualified myself by a course of hard study and attendance on the lectures of the most distinguished professors, my great object was to get employed in scientific voyages of discovery. Wearied and tired out with the buffeting about in small craft, ofttimes very uncomfortable vessels, employed upon unhealthy stations, wanting in interest and influence to obtain for me the appointment to a frigate or other desirable ship, throughout the whole course of my service in the navy, I was yet compelled to serve, and, indeed, it was my duty so to do, and not only simply as a duty, but for time, a great desideratum with our particular class, with whom "service time" alone counts for either increase of pay or promotion.

Therefore I shall not dwell in detail on this anything but interesting period of my services afloat, but shall very briefly indeed transcribe from my somewhat voluminous daily journals, only a few passing events, as they occurred at the time, so as to enable me to keep up a somewhat unbroken chain of the chief events and reminiscences of a long and eventful life, that I may have the more space left me to record and do justice to those Polar voyages, both to the Arctic and Antarctic Seas, in which I have had the good fortune to be engaged, and which have ever been the stirring ambition of my life. Indeed, but for this feeling, my pen would not have been taken up now as an octogenarian.

The subsequent three years from which I have to record were spent in two small miserable craft, and for the most part on my old station, the West Indies, where I had already suffered so much from the climate and other depressing influences; which I can only look back to with unavailing regret, as so much time, health, and energies utterly wasted.

The first appointment was to a small surveying ten-gun

brig fitting out at Plymouth, dated July 9th, 1831, which I joined on the 20th, and on November 22nd I attended a survey at Plymouth Hospital, and dined with Dr. Armstrong, the surgeon of the hospital, meeting at dinner my old friend and brother officer, Dr. McArthur, the surgeon of the *Fly*, of eighteen guns, fitting out here.

CHAPTER VII.

Voyage to South America—Cape de Verde Islands—St. Paul's Rocks—Bahia—Rio Janeiro—Return.

Tuesday, 27th.—After six months had been dawdled away in fitting a small ten-gun brig for sea, and after several unsuccessful attempts to get down channel, and as often driven back by bad weather, we at last got under weigh from Barnpoole, and stood out to sea.

Monday, 16th of January, 1832.—We anchored off Porta Praya, in the island of St. Jago, Cape de Verde Islands, and I landed there. On the 18th I paid a visit to the remarkable old baobab-tree, an African tree growing in an open space to the westward of the town. On a second visit to this tree, on the 20th, I shot off the only blossom from its summit with a ball from my old double-barrelled gun, and it fell to the ground uninjured, the flower as large as that of the *Magnolia grandiflora*, with fine white fleshy-waxy petals. I had ascended the tree previously, but could not get within reach of it, and, as a memento, I cut my initials, with the date of the year, high up the main stem.

On the 23rd I landed upon a shooting excursion, and descended the steep, perpendicular sides of a deep ravine to the eastward of the town and opening to the sea. Numerous tropic birds (*Phæton ætherius*) were hovering over the surf-beaten rocks at the entrance, with a wild cat at the bottom, altogether presenting a wild and striking scene.

Thursday, February 2nd.—I went on shore again, and on measuring the baobab-tree, I found it $36\frac{1}{2}$ feet in circumference.

Saturday, 4th.—Made an inland excursion to the Valley of St. Domingo; shot a rock pigeon on my way to the village, nestled amid shrubs and trees.

Monday, 6th.—I paid another visit to the baobab-tree, and took a sketch of it. On Wednesday, the 8th, I took a sketch of the place, went on shore again, and at three p.m. sailed. During my stay here I shot several "land" kingfishers or "Jacamas," differing widely from the common kingfisher in its habits, being found in the bushy, dry valleys where there is no water, and living on locusts and other insects. Its bright blue and black plumage sparkling in the sun's rays as it flits from branch to branch.

Thursday, 16th.—Hove-to off St. Paul's Rocks in the middle of the Atlantic, and took a sketch of them. On the 19th, at midnight, anchored off Fernando Noronha, and sailed again next day.

Tuesday, 28th. Eleven a.m.—Anchored off Bahia, found H.M.S. *Samarang*, Captain Lord William Paget, lying there. Dined on board of her with the captain and ward-room officers, and again on March the 3rd and 8th. Dr. George Todd Moxey, who entered the service about the same time as myself, was surgeon of her.

Saturday, 10th.—I went on shore to the Convent of Solidada, and purchased of the nuns some of the feather flowers for which this place is so famous, and on Sunday, the 18th, we sailed; 29th, anchored off the Abrolho Shoals, and I landed on one of the islands, where I caught a young tropic bird in the long grass. Sailed on the 30th.

Wednesday, April 4th.—Anchored in the harbour of Rio Janeiro, and landed on the following day.

Thursday, 12th.—Walked to Boto Fogo, where there

is a fine white sandy beach, and paid a visit to the Botanic Garden. On the 26th I crossed over to Praya Granda in one of the passage-boats.

Saturday, 28th.—I went to the market-place and purchased a grey parrot; and joined H.M.S. *Tyne*, of twenty-eight guns, Captain Charles Hope, for a passage to England. Went on board about midnight. Having found myself in a false position on board a small and very uncomfortable vessel, and very much disappointed in my expectations of carrying out my natural history pursuits, every obstacle having been placed in the way of my getting on shore and making collections, I got permission from the admiral in command of the station here to be superseded and allowed a passage home in H.M.S. *Tyne*.

On Sunday, 29th, at day-break, we sailed for England, and the *Tyne* proved to be just as comfortable a ship as the one I had just left was the very reverse. We had a fine passage home; and had on board as a passenger a lieutenant in the French navy, of the name of Le Jeune, a young man of most prepossessing appearance and manners, amiable and intelligent, who had been engaged in one of the French scientific voyages of discovery round the world, under Admiral Duperrey, in the *Coquille*. He had just left one of the ships of the French squadron on the station to proceed to France, to take up an appointment on the home station. He and I very soon became friends, attracted to each other by the mutual interest we felt in natural history, geography, and other branches of science. My brother officer, the surgeon, Peter Cunningham, was a brother of the well-known Allan Cunningham, writer and poet, and was himself an author; so that altogether we formed a very happy mess, the most so by far of any ship I have ever belonged to. Captain Hope himself was an excellent officer and a kind-hearted, sociable man, and had been a midshipman with the late

Sir Murray Maxwell in the loss of the *Alceste* frigate in the China seas. We had scarcely been three days at sea before I received an invitation to dine with him, on the 1st of May. Early on the morning of the 10th we passed the lone island of Trinidad, and I took a sketch of it.

On the 12th I again dined with Captain Hope. On the 21st we crossed the line, on the 26th I dined with the captain, and on the 31st crossed the tropic.

Friday, June 1st.—The north-east trade-wind still blowing fresh and squally, and we passed a great deal of the gulf-weed.

Monday, 4th.—The ship's company were exercised at the great guns, firing at a mark. On the 8th we fell in with a wreck, the schooner *Constance*, of Cherbourg, water-logged, with only her fore-mast and bowsprit standing, her decks awash with the sea, and covered with barnacles. Consequently she must have been drifting about for a long time.

Wednesday, 13th.—Strong breezes and fair. Passed a large ship outward bound, and I dined with the captain.

Saturday, 16th.—Made the land. We were surrounded by vessels of all sorts, and saw the Lizard Lighthouse. At two p.m. we were in sight of the Eddystone, and passed Plymouth.

Monday, 18th.—At six a.m. anchored at Spithead, saluting whilst under weigh. At eleven a.m. I landed at Point, reported myself at the admiral's office, and took up my quarters at the old "Star and Garter."

On the 19th the *Tyne* sailed for Sheerness, and I went to the custom-house for my baggage. On the following day I called on my old friend Lieutenant Rogier, R.N., at his lodgings in Green Row, Southsea, and stayed to tea with him.

Thursday, 21st.—At ten a.m. I left by the "Rocket"

coach for town, reaching my old lodgings, 15, Duke Street, Adelphi, at seven p.m., where the celebrated African traveller, Captain Clapperton, R.N., once lodged.

Friday, 29th, one p.m.—I attended the invaliding survey at the Admiralty, and found myself again on half-pay.

CHAPTER VIII.

On half-pay—With the combined squadrons in the blockade of Holland —Narrow escape in the Downs—Aground on the Goodwin Sands —Court-martial.

Monday, July 16th.—This evening I went to a party at Greville Place, Kilburn, to which I had received an invitation from the mother and sister of my late messmate, Augustus Earl, the artist and celebrated draughtsman of tropical scenery and savage life.

On the 23rd I saw the new library of the British Museum for the first time.

On Friday, the 27th, when walking through the department containing Captain Cook's collection from the South Sea Islands, the keeper of the room called my attention to a remarkable-looking old gentleman dressed in the costume of the last century, wearing a very broad-brimmed white hat with a black band round it, and a square-cut broad-skirted coat, and waistcoat of similar dimensions, who, he said, was an habitual, almost daily, visitor there, and no less a personage than Colonel Molesworth Phillips, the marine officer who so gallantly defended Captain Cook when he fell beneath the dagger of a Sandwich Islander, in the ever to be lamented affray with the natives of those charming islands. Poor old Phillips after all fell a victim to the cholera epidemic at the time so general in the metropolis.

Being desirous of seeing some of thc worst cases, I on

the 3rd of September called on an old brother officer, Peter Cosgrave, Surgeon, R.N., practising in Surrey Street, Strand, and accompanied him in his morning rounds through the wretched neighbourhood of the poor courts and alleys of Drury Lane, where some of the very worst forms of this dire and too frequently fatal attack met my eye.

On the 8th, finding agents' services did not justify the expenses incurred by employing them, I had the power of attorney cancelled, and dismissed altogether my agents, Messrs. Cook and Halford, of 41, Norfolk Street, Strand, whom I had originally employed, chiefly from their having been not only the agents, but also the executors of my poor father.

On the 10th I crossed over the new London Bridge for the first time; and on the 13th went to Yarmouth by the *Courier* steamer, and anchored off the bar on the following morning at five a.m,, and walked to Southtown.

October 3rd.—I received an appointment to a sloop-of-war of eighteen guns, fitting out at Portsmouth, to join the combined squadron of English and French in the blockade of the coast of Holland. I left Southtown on the 11th, passing a day or two in town on my way to join her.

On the 16th of October I joined her at Portsmouth, and on the 8th of November, at two p.m., sailed for the Downs.

On the 20th, at one p.m., weighed and stood over to the Dutch coast, a low, sandy, hummocky shore, studded over with windmills, and numerous large flights of wild-fowl and gulls tended somewhat to enliven the otherwise dreary scene, amid the squally, thick, often foggy weather, and not unfrequent heavy gales of wind which the squadron cruising off the Texel and Gorée had to encounter. We passed in sight of the famed Camperdown, off which Lord Duncan gained a victory over the Dutch fleet.

We passed January 1st, the New Year's Day of 1833, cruising between the Texel and the Scheldt.

Sunday, 13th.—We early this morning returned to the Downs; and when passing close under the stern of the French admiral's flagship the *Syrene*, a boat in charge of a midshipman from her was sent alongside of us with a message from the admiral's aide-de-camp, Lieutenant Le Jeune, wishing to know if his old friend and former messmate in the *Tyne* was on board, as it appeared he had recognized me on the poop as we passed astern of the *Syrene* in coming to an anchor. On the following day he came on board to see me.

Sunday, February 17th.—My friend Le Jeune dined on board with me in the gun-room.

On Friday, the 22nd, having landed at Deal, at three p.m., in the small dingy, to get some things for the mess, with the gun-room steward, and only four boys to pull the boat, accompanied by Webb, one of our midshipmen. As we attempted to get off again, the wind and sea had in the meantime increased so much as to swamp our boat, turning her bottom-up in the surf, through which we had to scramble on shore with a thorough ducking; and, having righted the boat, by emptying out the sand and water she had taken in, made another attempt at launching her through the heavy rollers and surf breaking on the beach. Things looked so unpromising that the Deal boatmen, sturdy fellows themselves, tried to dissuade us from making another attempt, urging with good reason too, that with a strong tide and wind against us, in such a nutshell of a boat, with a dark night before us, to reach the ship seemed hopeless. However, as I told them I had made up my mind to get on board, all I asked of them was to lend a helping hand by giving us a shove through the surf, which they at once willingly did, and I distributed a gallon of beer amongst these fine hardy fellows, and we succeeded this time in getting clear of the rollers setting on the beach. A fearful struggle with the elements we had at eight p.m., after three hours' incessant toil at the oar and helm, in which the midshipman and myself took turns alternately at the oar or steering. Our feeble boat's crew, early succumbing to the cold, wet, toil, and exposure, lay at the bottom of the boat as so much ballast. In the darkness of the night our only guide to steer by was the lights on board the ships of the squadron, densely black clouds and squalls passing over us, every moment threatening our destruction by swallowing up the frail skiff as she laboured against both wind and tide, tossed about on the fast rising sea like a nutshell. The first ship we made through the surrounding gloom was a Frenchman, and my companion was very naturally anxious to secure the first chance that offered, with the prospect of the Goodwin Sands before us, should we miss the ships of the combined squadron. I confess I very reluctantly allowed this chance to pass by; but observing another light in our course, which might prove to be one of our own ships, on board which I felt it would be far more desirable to seek a shelter for our boat's crew than on board a foreign one, I shaped our course for the next light, which happily proved to be one of our own line-of-battle ships, the *Malabar*, of seventy-four guns. But as we pulled under her quarter, we risked being knocked to pieces against her huge sides, by the swell of the heavy sea which swept past them. We had already drifted past the gangway before we could secure the rope thrown to us, almost buried in the waters surging alongside. When under the forechains, I most fortunately caught the rope, and, clinging to the thwarts of the boat with my feet and legs, I held on with a firm grasp for dear life, till we had all succeeded in clambering up the steep sides of the vast hull, pitching and rolling in the heavy sea that was running. The boat was then hoisted on deck, and the ward-room officers, after we had changed our wet clothes, took my companion, the midshipman, and myself down to the mess-room, where we sat down to a comfortable warm supper; after which, as the commander was on shore sick at the hospital, I had his cabin allotted me to sleep in for the night; and my boat's crew were well taken care of and made comfortable. A signal was made to my own ship that we were all safe, and the next morning, between seven and eight, a boat took us on board, having to pull against both wind and tide. A buoy-rope aided us on board at nine a.m.

On one of our many cruises in the North Seas, whilst blockading the coast of Holland during this tempestuous winter, I landed on the 6th of March at Ostend, and took a survey of the town. On the 13th, at 1.30 p.m., the combined squadron, with the *Donegal*, bearing the flag of Sir Pultenay Malcolm, and the French flag-ship *Syrene*, sailed for the coast of Holland, returning again to their anchorage in the Downs on the 19th.

On the 20th, at one p.m., I went on board the *Syrene*, to return the visit of my friend Le Jeune, the admiral's aide-de-camp, who was busy getting the ship under weigh for Cherbourg. But he showed me round the admiral's cabin and main-deck, when I returned on board.

Tuesday, April 2nd.—Having got under weigh from the Downs for a cruise, when off Calais, on the 4th, we captured a small Dutch galliot, laden with fruit, and took her in tow, when, thick and hazy weather coming on, we somehow or other got out of our reckoning, and, much to the astonishment of all, found ourselves all at once hard and fast on the Goodwin Sands. It occurred about the middle of the day. I was seated at the gun-room table at the time, and felt the shock, immediately followed by a second, and, upon going on deck, found all in confusion there, with an entire absence of that coolness, self-possession, and seamanship so much needed in the commander in such an emergency as the present. The paltry prize had been cast off, and most of the guns were thrown overboard without even buoying them, and those that remained were fired as signals of distress, and the ensign reversed. Between five and six p.m. I leaped out of one of the foremost ports on the starboard side, and walked round her, from the bows, round her stern, touching her copper sheathing. Astern of us a long line of breakers extended as far as the eye could reach. The South Foreland Light on the starboard bow, the North Foreland and Gull Stream Lights on the starboard quarter, and astern of us the town of Deal, and the combined squadron at anchor in the Downs, and on her port and weather side the open sea.

Just before dark the whole sky became overcast with a densely black cloud, threatening a stormy, tempestuous night for our destruction. But fortunately for us all passed off with a squall of wind and rain. At the time we struck, about an hour and a half after high water (spring tides), we had topgallant sails set. Her starboard side rested on the sands, and as the tide made she righted rapidly and in the most gentle manner. We had the moon nearly ahead of us as she cast round going off. Boats from the combined squadron, and from Deal beach, came to our assistance, and on the tide making, about ten p.m., we fortunately got afloat again, and anchored off the North Foreland Light.

On the following morning we weighed, anchoring again at two p.m. in the Queen's Channel; having only the two long guns forward, and three carronades remaining on deck. The ports had been stove and knocked out in getting the others overboard, with several casks of provisions, and all the water started from the tanks, together with a top-mast, which, singularly enough, we had picked up. It had belonged to a ship wrecked on this very spot not long before, the hull of which lay embedded in the sand a hundred or two yards inside of us, whilst our ill-starred ship lay high and dry on the hard sand. We were taken into dock at Sheerness. I went on duty to Chatham to attend a survey, and lunched with an old acquaintance, Adamson, the assistant-surgeon of the yard. We received orders to proceed to Portsmouth for a court-martial on the commander, who was sentenced to be reprimanded, and admonished to be more careful in future. The president was Sir Frederick Maitland, of H.M.S. *Victory*. This took place on Thursday, the 18th of April.

On the 27th, finding that our ultimate destination was the West Indies, a station on which I had already suffered so much in health and every discomfort, it may be concluded I was in no disposition to return if I could avoid it, more especially in a ship from which both lieutenants were superseded at their own request. She subsequently, in the latter part of her commission, became so unpopular on the West India station as to be put in "Coventry" by the military garrisons of the various islands, by whom she was considered and looked upon with as little prestige as a privateer craft might have claimed.

Consequently, on the 29th I wrote a letter to the Admiralty, requesting to be appointed to some other ship, and had an interview with the admiral there on the subject, which led to a second application from myself, on the 3rd of May, equally unsuccessful, and leaving me no other alternative than to sail for the third time to a station which destiny seemed to have marked out for me to be the very bane of my existence.

May 10th.—We got under weigh for Falmouth,

anchoring there on the 14th. I went on shore at five a.m. on the following day, and breakfasted at "Pierce's Hotel," returning on board at eleven a.m. My old ship, the *Hyacinth*, was at anchor in the harbour.

A water-spout seen between Bermuda and Jamaica on August 7th, 1833.
(*See page* 234.)

The Moro Castle, Havana. (*See page* 243.)

CHAPTER IX.

Third voyage to the West Indies—The yellow fever—At Havanha—Invalided home.

May 15*th*, 1833.—At noon sailed from Falmouth; and Thursday, the 20th of June, made the island of Deseada, passing in sight of Antigua, and between the islands of Guadaloupe and Montserrat. All these islands were in sight at the same time. On the 25th sighted St. Domingo, and on the 27th at six p.m. anchored in Port Royal Harbour, Jamaica. On the next day I landed, and had the famed "porter-cup" at Johnny Ferong's, and went to Kingston. My old friend Dr. McArthur came on board to see me.

Saturday, 29*th*.—Six a.m., sailed for Bermuda. Cleared the windward passage with a fine breeze, sighting Cuba, Cape St. Nicholas Mole, Fortune, and the Crooked Islands.

Friday, July 12*th*.—Hove-to off George Town, Bermuda. Received despatches for the admiral at Halifax, and at six p.m. made all sail again. We are now fairly in the Gulf Stream, the colour of the sea having changed from the bright blue of yesterday to the dark green of to-day, and surrounded on all sides by the gulf-weed. Stormy petrel numerous; a hundred or more seen at the same time.

On the 15th strong breezes. Going nine knots. The water returning to its blue colour, and the gulf-weed disappearing. The evening very chilly, the temperature next day falling to 60° Fahr.

Wednesday, 17th.—Nine a.m., saw Sambro Light, York Redoubt, &c., and the pilot came on board. Saluted the admiral's flag, and about noon anchored above the dockyard, opposite to the admiral's house, having passed McNab's and the St. George's Islands on the starboard side. Found the *Vernon*, flag-ship, at anchor here. At four p.m. I landed in the jolly-boat at the Victualling Office. Had some of the wild strawberries and cream at Jones's, the confectioner's, and took a stroll round the town.

Thursday, 18th.—I went on shore again this morning, and at five p.m. we sailed for Bermuda, where we arrived on the 27th, passing through the Narrows and Murray's Anchorage to the Ferry Anchorage, which we reached at ten a.m., and I landed at George Town at one.

Tuesday, 30th.—At six a.m. I landed at the ferry near the bridge, in the pilot's boat, on a shooting excursion, extending our excursion to the westward some four miles or so. Had a glass of milk at a house on the road, where we saw a fine young eagle, said to have been hatched over General Washington's tomb. I gave him a bird I had shot, for his lunch. Saw some tropical birds in Shelly Bay, and after calling at old Tucker's, the pilot's, for some sea-rods, reached George Town at seven p.m., and returned on board.

Wednesday, 31st.—We received a company of the 8th Regiment on board for a passage to Port Royal, and sailed. The birds I shot at Bermuda were fifteen blue and two red birds, four ground-doves, five chick-de-willock, and four black or cat-birds.

Wednesday, August 7th.—About midday I saw a

waterspout, appearing first like a dark, then changing to a light, column. On the 14th we cleared the Mona Passage, passing in sight of Zacheo, Mona, Monica, and Porto Rico. On the 20th sighted Jamaica off the White Horses and Yallah's Hill, at three p.m. on the following day anchored in Port Royal Harbour, and the next morning disembarked the troops.

Monday, 26th.—Went on shore at five a.m. in the cutter, to Green Bay, and shot several birds; returning on board at nine a.m., when I attended a survey on board of the *Magnificent*. At six p.m. I dined at the artillery mess, Port Royal, and wore our new full-dress uniform for the first time; returned on board at nine p.m. from Johnny Ferong's wharf.

Friday, 30th.—I landed at six a.m. on a shooting excursion to the Palisades; shot several small birds and a John Crow, or carrion vulture, and returned on board at ten a.m.

Tuesday, September 3rd.—We sailed for Chagres and Nicaragua, on the Spanish Main, and during the remainder of the time I remained on the station we were chiefly so employed—carrying the mails from Port Royal, &c. So that I shall not enter into a dry detail of the dates of arrivals and departures, and only note the day of the occurrence of anything worth recording in these short cruises.

On the 9th the ship was literally in an atmosphere of butterflies. In the forenoon I caught a shark astern nearly six feet in length, with a small "sucker-fish" attached to him. A slice of him, cooked with port wine sauce, for dinner, proved by no means unpalatable.

On the 11th, when off Chagres Castle, we encountered a violent storm of thunder and lightning, accompanied by torrents of rain, like a deluge. The thunder in loud crashing peals, and the lightning most vivid. One flash struck the ship, but without any damage.

On Monday, 16th, having anchored off Nicaragua, at

four p.m., I landed and took the mail to the Governor's, extending my walk along the beach to the entrance of the river, over which I crossed to the island, through a whole squadron of sharks and alligators, rising close to the boat on all sides. I landed on Sandy Point, where our cutter's crew had made a blazing fire, and were employed hauling the *seine* for fish. I returned on board at eight p.m., and on the following day we sailed for Salt Creek. When hove-to with the mail off this place, on the 19th, I shot at and wounded two sharks; one of them, ten feet in length, having a ball through the neck, was subsequently caught. On returning to Chagres I waited on the Governor on the 23rd, and made another shooting excursion or two; had a fine clear view of the Cordilleras, one of the peaks appearing to emit smoke, as if from a volcano. After returning to Port Royal, we again sailed for the Main; this time for Carthagena. Having sent the mail on shore on the 10th, on the following morning we passed through the Boca-Chica, between two forts, and anchored in the harbour on the 12th. I landed on the island on a shooting excursion, and bagged twenty-two birds.

On Monday, October 14th, six a.m., I landed at the market-place upon another excursion. Ascended the celebrated mountain of La Popa, so well described by Smollett in his "Roderick Random," skirting the hill, through the bushes, to the summit. At nine a.m. I entered the monastery of Nivestra Senhora de la Popa. Here we had a bottle of claret, and some orange cordial, and I shot a shrike and a ground-dove, within its walls, for my ornithological collection, and as a memento of my visit. Returning by the main road, we fell in with some wood-cutters felling trees in the forest, whom we much amused by firing at a mark with our guns. Got some cocoa-nuts at a hut, and after wading through a lagoon returned on board at six p.m., I having shot twenty-two birds.

Returned from the Main to Port Royal Harbour on the

23rd. On the 26th attended an invaliding survey on board the *Magnificent*, and afterwards went to Fort Augusta, and dined at the mess of the 8th Regiment, in full-dress uniform. Slept at the fort, and next morning breakfasted with one of the officers at his quarters; a second breakfast at the mess-room at one, and returned on board at three p.m.

Saturday, November 2nd.—Paid another visit to the 8th Regiment at Fort Augusta; breakfasted at the mess, and afterwards joined a party of the officers in their boat, the *Black Duck*, on a shooting excursion up the river, through a morass overgrown with rushes, &c.; and after wading through black mud and swamp, toiling in the heat, did not succeed in getting any wild-fowl. I only shot two black-plumaged birds; one from a flock flying overhead. After pulling through a mangrove creek, we got back to the fort at six p.m., and I dined at the mess. On the following morning, Sunday, I breakfasted with the officers, and afterwards visited the Military Hospital, returning on board at two p.m. in their boat the *Black Duck*. Sailed again on the 11th for Chagres and Nicaragua, and landed at each place.

On Friday, November 22nd, when on the passage between the above places, in the evening, a fine osprey was hovering over the mast-head, and ventured to alight on the weather-end of the main-topgallant yard, when I shot it from the poop and it fell on the quarter-deck. I at once skinned and preserved it. Its dimensions were: Extent of wings, five feet; length, twenty-two and a half inches; circumference, fifteen inches; eyes, bright yellow, with faint brown reflections; tarsus, pale blue. On the following day we had the osprey cooked for dinner, and found it far from unpalatable. Several others seen.

Saturday, 30th.—Having anchored on the previous evening off Nicaragua, I went on shore before daylight this morning on a shooting excursion, landed at Point Arenas, and walked along the coast for about four miles, and was fortunate enough to add a pair of the beautiful tiger bittern to my ornithological collection; one bird falling to each barrel, right and left, as they rose from a mangrove swamp (male and female). I also shot the pretty "Parra jacana," and a parrot. Most of the birds seen were white egrets, in flocks, with kingfishers, sandpipers, and a few ducks and parrots. Iguanas clinging at full length to the topmost branches of the trees, where-ever there was water underneath for them to fall into, beneath which they dived for security, whenever alarmed and disturbed from their look-out. Returned on board at eleven a.m. At two p.m. we got under weigh for Salt Creek, where I also landed.

December 16th.—Returned to Port Royal, where the *Vernon*, bearing the admiral's flag, was lying, with orders for us to prepare for sea again; and at daylight on the 19th sailed for Port Antonio, at the east end of Jamaica, anchoring there on the following day at four p.m.

Saturday, 21st.—Dined at the mess of the 37th Regiment, and on the 28th I breakfasted with Gregory Johnstone, at his estate, Anchovy, Richmond Hill. I learnt from him that he was one of the claimants to the Annandale peerage, and had been the owner of Fritton Decoy, in Suffolk, where he once resided, a spot so well-known to myself from having passed a year of my early youth at the village of Belton in its vicinity. And although he was of a retiring disposition, and lived almost the life of a recluse on his estate, this circumstance, reviving old associations of the past, soon brought us together on the most friendly terms. I rambled over Johnstone's grounds, and shot seven birds, returning on board at six p.m.

On Tuesday, 31st, I again breakfasted with Johnstone at Richmond Hill, and dined with him at Anchovy; riding to both places. We spent the evening where we breakfasted, at Richmond Hill. I shot three mocking-

Birds (*Turdus polyglottus*), and a John Crow (*Vultur aura*).

Thursday, January 2nd, 1834.—I went on shore and called on Dr. Arnold and Gregory Johnstone, with both of whom I had dined several times, and they both dined on board with me to-day.

Saturday, 4th.—I received an invitation from the commanding officer of the 37th Regiment to dine at the mess to-morrow, and from Gregory Johnstone to breakfast with him at Richmond Hill, and to dine at Anchovy on Monday next.

Monday 6th.—Went on shore, and to Richmond Hill; accompanied Johnstone round his grounds with our guns. He shot the little bittern (*Ardea minuta*), and gave it to me for my collection, and I skinned it at once. It rose from a mangrove swamp. After dining at Anchovy we visited his sugar plantation in search of snipe, these birds frequenting the furrows filled with water. But being late in the day, and no wind, they were so shy that I only got a shot or two, and missed them. After dark we rode to the barracks, and I returned on board in the 37th's boat to dress in full uniform, for the ball given by the townspeople to the naval and military at nine p.m. I had a long and interesting conversation with Johnstone in the ball-room, whilst the dancing was going on, as we were neither of us dancers. Returned on board at two a.m., and sailed for Port Royal.

Monday, 13th.—We got under weigh for Black River, anchoring there at two p.m. On the following day I went on shore, and brought off with me a detachment of the 8th Regiment, all down with yellow fever in its worst form, to convey them to the hospital at Fort Augusta. Weighing at daylight next morning, we took on board another detachment from Old Harbour on the 17th, and at daylight on the following morning sailed for Port Royal, where we landed the troops the same night.

Some died on board, and many others subsequently in the hospital.

On the 23rd I dined at the mess of the 8th Regiment at Fort Augusta; and on the 25th, at daylight, we sailed for St. Jago de Cuba, anchoring there on the 28th at four p.m.

Thursday, 30th.—I went on shore on a shooting excursion. The country in the vicinity was very beautiful. Some portions of the undulating hills, like a green lawn studded with clumps of forest trees, recalled to mind a country park in England. We returned on board at four p.m., over-heated and fatigued, after wading through swamps in search of birds.

February 4th.—Sailed for Port Royal, which we reached on the 6th, and proceeded to sea again on the 21st for Carthagena: on the 24th anchored there. On the 28th, about dusk, when at anchor off the Boca-Chica, vast numbers of a small species of bat passed over the ship in one continuous stream, so closely packed there must have been myriads of them. They proceeded from the direction of the fort, across the harbour.

March 1st.—Got under weigh for Port Royal. Whilst at anchor off the Boca-Chica, on the 28th, I caught one of these bats as it flew over my head whilst I was walking the quarter-deck, which I preserved, and found that it was with young.

Wednesday, May 14th.—We sailed for the Havanha, a place I had long ago wished to see, but an opportunity for so doing had never hitherto offered; the last two months having been occupied in the very monotonous service of carrying the mails backwards and forwards between Port Royal, Chagres, Nicaragua, Salt Creek, and Carthagena, together with surveys at Port Royal Hospital. On the 17th we sighted the Isle of Pines, a noted resort of pirates and slavers, and on the following day we rounded Cape Antonio.

Friday, 23rd.—We anchored off the Moro Castle, and at sunset warped into the harbour of Havanha. On the following day, at nine a.m., I landed for the first time in this fine city, certainly the metropolis of the West Indies, and replete with associations carrying us back to the time of Columbus, whose remains repose in the cathedral here. After taking a hasty survey of the city itself, I made an excursion to the Passeo, or public promenade, the Rotten Row of Havanha, returning in a *volanté* to Custom-house Quay, when I dined at Madame Roulliet's Hotel, and got on board again at six p.m. The day was cloudy and showery.

Sunday, 25th.—I paid a visit to the cathedral, where I saw the portrait and bust of the great Columbus, whose bones are preserved in a box placed in a niche of the wall to the left of the altar as you face it. Afterwards I rode out in a *volanté* to the Passeo, and walked through the Botanic Garden. On my return I had coffee, and one of those most deliciously cooling and refreshing draughts we called "iced grapes," at the General Café.

Tuesday, 27th.—I paid the celebrated Cabana's cigar manufactory a visit; when he himself very kindly showed me round the whole establishment, and whilst talking with him, he made me up with his own hands a specimen cigar from one of the finest leaves he could select, and which, after the lapse of half a century, I have still in my possession, in my desk, not being a smoker myself. Returned on board in the afternoon.

Wednesday, 28th.—I landed at Regulus, in a passage-boat, on a shooting excursion in the suburbs of Havanha, which I reached about noon. Having walked round the shores of the bay, I took a *volanté* through the town to the landing-place at the dockyard. I shot two small hawks and a finch. Passed a number of mango-trees, laden with fruit so ripe that they fell with the slightest touch to the ground, which beneath was strewed with them. I gathered and ate several; but this handsome, tempting-looking fruit does not always justify one's expectations, as some varieties have a very unpleasant turpentine flavour. I got a refreshing draught of cocoa-nut milk at a cottage. From the highest range of hills, crested with a row of the tall, erect, and elegant cabbage-trees seen bounding the horizon from the harbour, I had a magnificent view of the richly-cultivated valley beyond, and a commanding prospect of Havanha itself.

I passed through several plantations of tobacco, capsicum, sweet potatoes, melons, plantains, &c., and through one or two pretty gardens. Only saw two or three pigeons.

Thursday, 29th.—Eight p.m., I went to the opera: left at nine, and returned on board at ten p.m. After we had left we discovered that we had made a great mistake; having left at the end of the first act, under the impression that it was all over. The audience all left the house with tickets of readmission, which we had not been aware of at the time.

It is a spacious building, well lighted up by a handsome chandelier and two tiers of lamps suspended from the centre of the roof, which was ornamented with figured paper, yellow and salmon being the prevailing tints, relieved by a dark neutral shade. The three tiers of boxes were surmounted by a gallery, and in the pit were about a dozen rows of seats, one or two of them being covered with red leather in each row. Over the stage was some motto, and an imitation scarlet festooned curtain. The *prima donna* was a tall, showy woman, with dark hair; over which she wore a white cap, with the long ribbons flowing loose over her shoulders; a white gown trimmed with blue ribbon, and short sleeves; a blue silk apron, white stockings, with shoes having sandals of black ribbon round the ankles; dark bracelets, like hair, round her wrists completed her costume. She had fine, expressive eyes, full of animation, and a well-modulated voice; age about thirty.

Monday, June 2nd.—I went on shore and took a survey of the city, returning on board at five p.m. On the following morning, after breakfast, I went on shore again, visited the café, and crossed the harbour to the Moro Castle, inspecting the platforms and the fortifications from the lighthouse to Fort Cabanas, returning on board about two p.m. Next day we sailed at daylight on a cruise to the westward along the land. On the 10th, at five p.m., returned to our anchorage at Havanha, and on the 14th I went on shore, returning on board again in the afternoon.

Sunday, 15th.—Sailed from Havanha at daylight for Jamaica. On the 23rd a large species of ardea (*Ardea scolopacia*) was caught on board, in sight of the Isle of Pines, and sent down to me by the officer of the watch. Its movements were peculiar and grotesque; curving its very long, slender neck, and bending down its head, made low bows, then suddenly starting away, with its head projecting and half-unfolded wings, it would strike the deck with its long, curlew-like beak, at the same time making a kind of half-hissing, half-blowing sound, then all at once quietly commence pruning its feathers; the general gait being stooping. On the 26th two more of these singular birds were caught and sent down to me by the boatswain.

Monday, July 7th.—After a long beat-up from Cuba we anchored at ten a.m. in Port Royal Harbour, where orders were awaiting for us to prepare for sea again.

Having again suffered from the yellow fever, and the effects of this climate, I went into Port Royal Hospital as a patient on Wednesday, the 9th of July; and finding my general health very much broken up, and thoroughly disgusted both with the ship and the station to which I had been appointed, without any other alternative, on quitting England, I at once resolved to invalid for a third time; and accordingly applied for a survey, which was held upon me at the hospital at eleven a.m. on the following day, and I was ordered a passage home, "for the preservation of my life," by the first packet. I took with me my pet Isle of Pines ardea, alive; but on giving it in charge to the assistant-surgeon of the hospital, he placed it in the verandah, where the ants killed it.

Wednesday, 16th.—About the middle of the day, the sky became suddenly densely overcast with black clouds, and some showers of rain fell. It blew almost a hurricane, and in the middle of one of the heavy gusts, a poor, unhappy mocking-bird (*Turdus polyglottus*), I had noticed patiently brooding over her eggs in a nest, built in the fork of a tree opposite to where I was in the habit of sitting in the verandah, had her nest, with its treasures, blown down, and the eggs broken and scattered on the ground beneath.

Saturday, 19th.—In the forenoon the *Skylark* packet arrived; and on the 25th I wrote a letter to the commodore for a passage home in her; and at four p.m. was discharged from the hospital on board the receiving-ship, *Magnificent.* Returning again on shore, I called on Dr. Linton, the surgeon of the hospital, from whom I had received the greatest kindness and attention whilst in the hospital. Between seven and eight p.m. the order for my passage home came on board the *Skylark*, and I slept on board of her.

Saturday, 26th.—The *Skylark* sailed for England on the 2nd of August, at ten a.m., and hove-to off Fortune's Island, and I landed there, but only for a few minutes. At two p.m. again hove-to off Crooked Island, and I again landed there, and went to the post-office. In the garden I discovered a humming-bird's nest, with two eggs, in a guava-tree, and I cut off the branch with the nest, and took it on board with me, after having had dinner and a stroll along the beach for about a mile, shooting a few birds. Made sail at sunset, and crossed the tropic next day.

Monday, August 18th.—This evening we experienced a very sudden and decided change in the temperature for the first time since quitting the tropics. The wind was easterly, and we all felt as chilly as if we had crossed the Arctic circle with the setting sun. We passed much gulf-weed and many porpoises, with a few stormy petrel and sheerwater, and a hawk's-bill turtle was caught.

Sunday, 31st.—Saw the two lighthouses on the Lizard; and at four p.m. anchored in Falmouth Harbour. Accompanied Lieutenant Ladd, R.N., the commander of the packet, to Captain King's, the superintendent of the packet service; and at eight p.m. I went to my old hotel, Selley's "Green Bank," where I slept.

Monday, September 1st.—At six a.m. started by the "Defiance" coach for London, where I arrived on the following day at ten p.m., and slept at the "Ship Hotel," Charing Cross. On the 4th went to the Admiralty, and took lodgings in Northumberland Street, Strand.

September 5th.—Passed as an invalid at the Admiralty.

Monday, 22nd.—I called on my former messmate, Augustus Earle, at Greville Place, Kilburn, whom I found very ill indeed, never having recovered from the effects of his late voyage; and it was but too evident there was now no hope for him. On the 26th I saw the exhibition of Sir John Ross's late expedition at Vauxhall Gardens. On the 29th I left No. 35, Northumberland Street, and slept at the "Spread Eagle," starting the following day at 6.30 a.m. by the "Star" coach for Southtown, where I arrived the same evening.

CHAPTER X.

Four years on half-pay—Pedestrian tours through England and Wales—Visit to Dublin—Appointment to the *Terror*—Ship paid off.

October 8th, 1834.—I left Southtown Terrace for London. On the 25th I took a lesson in bird-stuffing, having seen a case containing snipe in the window of my gun-maker, Wilkinson, of Pall Mall, so well set up, that I asked him for the address of the taxidermist.

On the 27th my friend and former shipmate, Captain Charles Gray, of the Royal Marines, the author of a small volume of poems, called on me, to ask me to accompany him round the great brewery of Truman and Hanbury. And so favourable an opportunity for seeing one of the largest of our metropolitan brewing establishments, next to Barclay and Perkins, was not to be lost. He introduced me to the manager, a friend and countryman of his, a Scotchman, who explained the whole process, and we tasted all kinds of the malt liquors from the casks, spent the day in a most agreeable and instructive manner, and obtained a thorough insight into the whole art of brewing.

Having just completed the period of service which advanced my half-pay from five shillings to six shillings per day, and finding that my state of health was in anything but a satisfactory condition, after having been compelled to invalid no less than three times from the West India station, it became a paramount necessity for me to obtain some respite for a time from such a harassing kind of service for its re-establishment. I therefore came to the conclusion that the most judicious course to adopt would be a line of action that would combine change of air and scene with occupation for both mind and body at the same time, and thus prepare myself mentally and physically for the satisfactory performance of my duties in any expedition of discovery and research it might be my good fortune in the future to be employed in, more especially as a naturalist and geologist, a hope which had not been altogether chilled or relinquished by the past uncongenial nature of the services which had hitherto fallen to my lot. To carry out my views I proposed to myself to make a series of pedestrian tours through England and Wales during the summer months; and to devote the winter season to courses of lectures on natural history, geology, botany, chemistry, and some of the professional courses, uniting with the attendance on these a course of reading at the libraries of the British Museum and the College of Surgeons.

I began my first tour in the winter, in consequence of a desire to remove my mother and sisters, at as early a period as possible, from Norfolk, a place so much out of my way that much expense and loss of time was incurred in going backwards and forwards to see them, myself being the only relative remaining to them since our father's death. This determined me not to allow the advent of winter to interfere

Monday, November 24th, 1834.—I started at 7.30 a.m. on my first pedestrian tour from my lodgings in Northumberland Street, Strand, with the main object now of finding a suitable place of residence for the removal of my mother and sisters to, having previously forwarded on my portmanteau by coach to Oxford, to be left at the office of the "Mitre Tavern" till called for. Thus, unencumbered by baggage, I now fairly commenced my journey from Hyde Park Corner at 8.20 a.m., dressed in a Florentine shooting-jacket and Swiss check trousers, with a change of linen in my pocket, compass and thermometer, road-book and note-book. I reached Maidenhead at five p.m., and took up my quarters at the "Commercial Inn," twenty-six miles from town, but, including the détour I made through Windsor Park, I had walked at least thirty miles.

In this, my first excursion, from which I returned on the 5th of February, 1835, I walked over 468 miles, through the counties of Middlesex, Berkshire, Oxfordshire, Gloucestershire, Herefordshire, Monmouthshire, Somersetshire, and Devonshire; visiting all the chief towns and places worth seeing; making a stay of a fortnight at Cheltenham, the most beautiful and pleasant of all our inland watering-places, as I think. I completed my tour at Plymouth; the last day's journey from Dartmouth, thirty miles, being, with the first day's to Maidenhead, the two longest days' work in the tour. I returned from Plymouth to London by the Dublin steamer.

Thursday, May 7th.—I left town at ten a.m. on my second tour, going through Deptford to Sevenoaks Common, where at 7.30 p.m. I took up my quarters for the night at the "White Hart Inn," a very comfortable inn, embosomed in shrubs and trees, and a pretty garden. My course lay through Kent, Sussex, and Surrey, along the southern coast to Brighton, &c. I was absent about eleven days, walking over 211 miles—the longest day's journey being twenty-seven miles—returning to town, through Kingston, on the 17th.

Monday, May 25th.—I left town on my third tour by the Edgeware Road; reached Tring, thirty-one miles, at 8.30 p.m., and slept at an inn there. In this journey through the inland counties to Liverpool—across the island from the banks of the Thames to the Mersey—I walked 227 miles in eight days, through the counties of Middlesex, Hertfordshire, Buckinghamshire, Northamptonshire, Oxfordshire, Warwickshire, Staffordshire, Cheshire, and Lancashire; the longest day's journey being thirty-eight miles. I entered Liverpool at nine p.m. on the 1st of June, and only remained there that night, at a very uncomfortable and dirty inn, although one of the largest in Dale Street.

June 2nd.—On going to the Steam-Packet Office this morning I ascertained that there was no direct conveyance to London without going *viâ* Dublin, and as two boats were to start in opposition to-day at two p.m. to meet the London steamer next morning, rather than lose a week I made up my mind to start at once. After seeing the Zoological Gardens, I embarked in the packet at two p.m. to-day for Dublin; and on Wednesday, the 3rd, after a fourteen hours' passage, I for the first time in my life set my foot on the shores of Hibernia, the land of my father's nativity. Having only time for a hasty survey of the place, however, I covered a good deal of ground before breakfast, which I had at the "Northumberland Coffee-house." Afterwards I walked through the Phœnix Park, about two miles distant; and at three p.m. started from the North Wall in the *City of Londonderry* steamer for London, where I arrived at seven p.m. on Sunday, the 7th, having put into Falmouth and Plymouth to land passengers on the way and get in coal. After an absence of about a fortnight I returned to my old lodgings in Northumberland Court, Nos. 2 and 35. At the latter I engaged additional rooms for my mother and sisters, whom I met on the arrival of the Yarmouth coach at the "Saracen's Head," Aldgate, at eight a.m. on the 19th of June, and at four p.m. removed them to No. 9, Craven Street.

Saturday, July 18th.—At 7.30 a.m. we embarked in the *Cornubia* steamer for Exmouth, where I had secured lodgings for them, and where we arrived at 5.30 a.m. on the 20th, and at eight a.m. took possession of rooms in Cleveland Cottage, a very pretty house in a good garden, for a month. During our residence there I made excursions to all the watering-places skirting the coast in the vicinity, to find a suitable place of residence for them. Lovely spots, and combining all the advantages of beautiful sea and land views, with a fine mild climate, and cheap living, which I should have preferred myself to any other part of England; but they could not make up their minds, coming as they did from a large town like Yarmouth, to take up a permanent abode at any of them, under the impression that the winters in such small towns would be too dull; and I had to proceed on to Plymouth with them, where a large bustling seaport town might offer a better chance of pleasing them. Accordingly, on the 18th of August, we all went to Torquay to catch the steamer for Plymouth, sleeping at "Cole's Inn," and on the following morning proceeded in the *Brunswick* steamer to Devonport. Took lodgings at No. 2, Home Park Place, until a suitable house could be found; when, on the 3rd of September, No. 2, York Place, Stoke, having been fixed upon, they took possession, and I furnished it.

This anxiety being off my mind, and seeing through the daily papers that my old shipmate with Parry, Captain James Clark Ross, was fitting out the *Cove* at Hull for the relief of the ice-bound whalers, I wrote to him on the 19th of December; and not getting an answer, I wrote again on the 23rd, offering to accompany him. On the 26th Captain Ross wrote to say that my first letter had never reached him, and that when he received my last, on Christmas Day, all the appointments had been filled up; but as the *Terror* was to follow him out as soon as commissioned, he promised to apply to the Admiralty for my being appointed to her. She is now lying here.

On the 5th of January, 1836, I made a personal application to the Admiralty for her, and on the 20th of February was appointed to the *Terror*, lying at Chatham. Soon after I had joined her there was a rumour on board that Ross on his return from the North would receive the command of an expedition to the Antarctic Seas for magnetic observations. This was on the 27th, and on the following day I dined with the Commander of the *Terror* at his lodgings on shore, to meet the celebrated nautical novelist, Captain Marryatt, whom I had never before seen. He talked in the same broad, humorous way in which he writes.

The missing whale-ships, with one exception, having found their way home, the Admiralty no longer considered it incumbent on them to send the *Terror* out; the consequence was our being paid off after having been some five weeks in commission; and on the 24th of March I again found myself on half-pay. On the 28th I returned to town, and on Good Friday, the 1st of April, I dined with my former shipmate, Charles James Beverley, and his wife, at their residence, Bethnal Green. They have a billiard-table in the house, and I played a game of billiards, for the first time in my life, under the instruction of Mrs. Beverley, who, like her husband, is very fond of the game. Beverley told me that he had also been a volunteer for the *Cove*, in search of the missing whalers, but like myself unsuccessful.

On the 5th I attended Professor Grant's introductory lecture on Fossil Zoology.

On the 19th I returned to Plymouth by the steamer, and on the 26th commenced a series of excursions in the vicinity, each occupying a day's journey, varying from twenty to thirty miles a day: to Plympton St. Mary, Ivybridge, Buckland Abbey (once the residence of the celebrated Sir Francis Drake), Tavistock, Crowndale (about a mile from the latter place was his birthplace),

Torpoint, Bickley Wood, and Saltash. In returning from the latter place, my favourite dog Hecla was nearly meeting with a serious accident. Whilst heedlessly chasing some small bird in his playful mood along the green bank bounding the Devonport lines, before he could arrest his speed he went over the precipitous edge of the fosse, down a depth of some fifty or sixty feet, and, what was scarcely to be anticipated, received not the slightest injury. He was a beautiful high-bred liver-and-white coloured spaniel, of most symmetrical form, intelligent and active, and was the constant companion of my daily rambles. The last of these excursions was on the 7th of July, over Maker Heights and Mount Edgecumbe to Cawsand Bay.

CHAPTER XI.

Fourth walking tour—West of England and Wales—Ascent of Cader Idris—A balloon on her Majesty's Coronation Day.

Friday, July 22nd, 1836.—I started, for the fourth time, from Stoke at 10.30 a.m., on a summer excursion, on my birthday, leaving by the Tavistock Road. Passed through the counties of Devonshire, Somersetshire, Gloucestershire, Worcestershire, Herefordshire, Shropshire, and Warwickshire. The first day across Dartmoor to Moreton Hampstead, thirty miles, where I slept at the inn. On the 30th I revisited my favourite watering-place, Cheltenham, where I remained until the 8th of August, at very comfortable lodgings. Visited the various spas; made an excursion to Tewkesbury and its old battle-field, and another to Malvern, ascending the ridge of hills above the Wells, and descending by the Witches' Pass, a wild, romantic-looking gap cut through the solid perpendicular rock. The 14th and 15th I spent in Ludlow; breakfasted, dined, and slept at the "Elephant and Castle," one of the oldest inns, having the rafters appearing black outside the walls, giving a magpie-like colouring. Had a fine view of the old castle from a lane called the Linney, on the Sunday evening, where I was caught in a severe storm of thunder and lightning, with heavy showers, the rain falling in torrents. Got shelter in a cottage there.

I reached Leamington on the 17th, and had intended

remaining here for a short time, for the benefit of the mineral waters. When finding through the daily papers that the *Cove* had returned to Hull from the North, and being anxious to see Captain Ross as soon as possible after his return, I resolved to start at once for town; so that my present tour was thus upset and brought to a premature close. I left Leamington by the coach on the following day, Saturday, the 27th, and arrived in London about seven p.m. I had gone over a distance of 353 miles on foot; the last day but one before my arrival at Leamington being the longest day's journey, thirty-four miles.

On Monday, October 3rd, I had an interview with Captain Ross at his lodgings, Westbourne Green, and volunteered my services for the next expedition, and he promised me that no one else should be appointed.

Tuesday, October 11th.—At eight a.m. I left the "Golden Cross," Strand, by the "Telegraph" coach, for Southampton, and on the 13th embarked in the *Saumarez* packet for the island of Jersey, where I arrived at noon of the 15th, on a visit to an old messmate residing there—at Bagot, about a mile from St. Helier's. During my stay, made six excursions in the island, going over from ten to twenty miles a day, making in the aggregate about ninety miles in all. Limited, but varied scenery.

On the 30th I left Jersey in the *Sir Francis Drake* steamer. We anchored off the town of St. Peter's Port, Guernsey, for an hour and a half, to land and to take passengers on board; which enabled me to see something of the island, by landing and making an excursion, as far as the limited time would permit. The day following we arrived at Plymouth, where I remained throughout most of the winter, with my mother and sisters, at Stoke, employing myself chiefly at the reading-rooms, and getting books from the library.

January 1st, 1837.—I attended the morning service

at the Dockyard Chapel. On the 3rd I called on Sir George Magrath, an old brother officer resident in Plymouth; and on the day following he called on me at Stoke.

On the 30th I embarked in the *Shannon* steamer for London, to attend the courses of lectures on Natural History, &c., during the winter season.

Monday, February 2nd.—I entered my name at the London University, to Dr. Grant's course of lectures on Zoology, and attended the first lecture at three p.m. The introductory one, however, had been delivered yesterday. The class was a very small one; the number of my ticket, about the last one, was twenty-eight. On the 3rd inst. I called at our office, Somerset House, and got a certificate for free admission to the various courses of lectures delivered at King's College during the session, half-pay medical officers of the navy having honorary admission on a recommendation from their Board.

From two to three p.m. I attended Professor Daniell's lecture on Chemistry, and on the same day, Dr. Todd's lecture on Physiology. On the 4th, from seven to eight p.m., Dr. Watson's lecture on Influenza. On the 6th, at eleven a.m., Dr. Royle's lecture on Materia Medica. From three to four p.m., Dr. Grant's lecture at University College; and Dr. Arnott's lecture on Surgery, at King's College, at seven to eight p.m. On the 7th, at ten a.m., Professor Partridge's Anatomical Demonstration, and afterwards Professor Rymer Jones's Microscopic Experiments.

21st.—Read for about two hours in the library of the Royal College of Surgeons, for the first time since it was opened, after being rebuilt. Began a regular course of reading of the best works on natural history, Humboldt's and others.

March 1st.—I visited the new museum of the college for the first time.

April 18th.—At four p.m. went to the theatre of the college and heard Stanley's lecture on the Bones.

Monday, May 1st.—Visited the rooms of the Royal Academy in Trafalgar Square, which were open to the public for the first time to-day, with the exhibition of paintings. Entered my name at King's College, to the courses of lectures on Botany and Experimental Philosophy.

Tuesday, May 2nd.—At 3.30 p.m. attended Professor Ritchie's lecture on Natural Philosophy, at the United Service Museum; at 4.30 p.m., Professor Owen's lecture at the theatre of the College of Surgeons, on Comparative Anatomy; and at eight p.m. the concluding lecture of the course on Compensative Anatomy, by Professor Rymer Jones.

Wednesday, May 3rd.—I attended Don's lecture on Botany, at King's College at nine a.m. On the 4th, I attended Professor Owen's lecture at the College of Surgeons, Sir Astley Cooper in the chair: it was crowded. The lecture embraced the history and progress of natural history, from the time of Aristotle and Pliny, with very graphic descriptions of the ancient Romans, spectacles of wild animals in the arena, &c. Altogether a most interesting and instructive lecture, replete with enthusiasm and eloquence.

Tuesday, 9th.—At eight p.m. went to Professor Wheatstone's lecture on Measures of Sound, at King's College.

Wednesday, 17th.—At three p.m. went to Professor Phillips' lecture at King's College, the introductory one, on Organic Remains.

The various courses of lectures which I had been in regular attendance on, since the commencement of the session, having now been brought to a close, and having nothing further to detain me longer in town, on June 1st I sent on my portmanteau by the Dublin steamer to

Plymouth, with the intention of making another pedestrian trip westward to Plymouth.

June 15th, 1837.—At seven p.m. I reached Stoke, on my fifth tour; having left Hyde Park Corner at 3.30 p.m. on the 1st; passing by Virginia Water, through Hounslow, Reading—by the celebrated Mary Russell Mitford's cottage at Three Mile Cross; continuing my course through Berkshire, Wiltshire, Dorsetshire (calling on Miss Anning and her mother, at their humble cottage in Lyme Regis, to see her fine collection of the fossils from the Lias Cliffs adjacent). Coasting Somersetshire and Devonshire, I slept at the "Valley of Rocks Hotel," beautifully situated above Linmouth, in the valley beneath, on the Linton Cliffs. When two miles from South Moulton I caught a young rabbit as it scrambled up the hedge-row from the roadside, and carried it home with me, passing through Tavistock to Stoke. During the fortnight I travelled over 335 miles. The longest day's work was thirty-four miles.

Saturday, July 22nd, 1837.—The anniversary of my birthday. I left Stoke on my sixth tour, and at one p.m. embarked from the Barbican in the *Sir Francis Drake* for Falmouth, which we reached at seven p.m. I walked on to Penryn; slept at the "King's Arms Inn" there; and on the following morning started for Penzance, visiting St. Michael's Mount, and making an excursion from Penzance to the noted Logan Rock and the Land's End: the view from the latter, as I stood at seven p.m. on a lovely summer evening on the edge of the overhanging cliff, was most charming. The Longships, a ledge of rocks, with its lighthouse about a mile distant; the wide expanse of the Atlantic spread out before me; and as I stepped on the extreme western point of England—a beetling rock, deeply cleft from the general mass—the sea beneath me smooth as a mill-pond, a fine hawk rose from the rock within a few feet of me, gliding downwards in a easy, graceful flight. The bird, whose back was towards me, had evidently been so intently occupied in observing something below the rock on which he was perched, as not to notice my approach until I was quite close to him. The loud and vociferous screaming of the numerous sea-mews, on being suddenly disturbed in their solitary haunt, might very possibly have for the moment distracted his attention. I now retraced my steps to Penzance, distant from the Land's End ten miles. Reached the "Dolphin Inn," where I slept, at ten p.m. The following morning early, I started for St. Ives, distant eight miles. From this place I took my passage in a boat for Ilfracombe, where I landed about midnight, and slept at the "Packet Hotel."

Wednesday, 26th.—Just as I had finished breakfast, a small sloop of seventy tons, called the *Happy Return*, with only two hands besides the captain, was about starting for Swansea, the place of my destination; in which I embarked, and in about three and a half hours was landed in Swansea, South Wales, at 3.30 p.m.

Passing through the counties of Carmarthen, Pembroke, and others along the coast of Wales, I visited Tenby (rounding St. David's Head), Cardigan, and Aberystwith, and entered North Wales by the Devil's Bridge, a cleft in the rock, or narrow ravine, four miles beyond which is a large posting-house, with the not very euphonious name of the "Gogerddam Arms," Ponterwyd, in a most bleak and lonely situation, surrounded on all sides by barren hills. The charges for very indifferent accommodation were exorbitant, as I have found to be generally the case with the inns in Wales, and far inferior in every respect to those in England.

Sunday Morning, August 6th.—After breakfasting at the inn, I started about noon for the town of Machynlleth, in Montgomeryshire, fifteen miles distant. One of the most toilsome, difficult, and perplexing routes I ever undertook.

My way lay over a bleak, barren moor, without a trace of road or path, tractless as the desert, with no other guide than my little pocket compass. My course being due north, over hills and hollows, through peat-bogs and valleys, till a mountain stream suddenly arrested my progress, when luckily I met a young man and woman, but who could not speak a word other than their native Welsh tongue, who by signs directed me to a hut, from one of the inmates of which I learnt that the stream was fordable a short distance further on. I was now about six miles on my way across the moor, from which my course lay over ridges of hills and a deep and broad valley, filled with peat bogs, through which I had to plod, with my feet well sodden with water. At last, on getting clear of the moorland, I entered a field of corn through a gateway, then followed a zigzag path through a deeply wooded glen, at the bottom of which a mountain stream dashed along in murmuring eddies through the channel it had cut for itself in the rocks, winding in serpentine curves through a perfect labyrinth of overhanging trees. This path I followed for about a mile, when I emerged upon the turnpike road about two miles from Machynlleth, situated in a valley, and which I entered about 5.30 p.m., thanks to my pocket compass; for without this I could never have threaded my way through such a succession of tractless bogs and hollows, hemmed in on all sides by rugged hills, before night had overtaken me.

Monday, 7th.—I left Machynlleth at eight a.m.; and at six p.m. reached the pretty rustic-looking little inn of "Penny-bont," or "Tally-y-lyn," grown all over with ivy. This was the most comfortable resting-place I had as yet met with in Wales; everything good, both attendance and food, and for my dinner I had excellent trout from the lake adjacent, and which nearly fills the lovely valley.

Tuesday, 8th.—At nine a.m. I commenced the ascent of the steep mountain of Cader-Idris, at the back of it, to make my way down the opposite side to Dolgelly. I reached the summit at 11.30 a.m., and at noon I began the descent; but where I attempted it the steepness and treacherous nature of the ground beneath my feet compelled me to edge round to a more promising part, where I accomplished the descent; and on my entering Dolgelly, at three p.m., I was told that even the natives here never attempted the spot I failed in. The distance I had walked to-day from Tally-y-lyn over Cader-Idris to Dolgelly was about twelve miles. On calling at the post-office I found a letter for me which upset all my plans for this tour, being nothing less than an appointment to the flag-ship at Malta, as a supernumerary, to go out in the next Falmouth packet to join her. As this letter had already been a week in finding me, I at once wrote to the Admiralty, explaining the cause of the delay, and made up my mind at the same time to return, *viâ* Dublin, by the steamer; for by so doing I should probably lose no more time than if I proceeded by coach, with the number of changes of coaches, and other delays on the road I should incur; and it would afford me an opportunity of completing a larger portion of my original intended tour. After sleeping at the "Ship Inn," I started on the following morning, Wednesday, the 9th, for Barmouth, ten miles distant, along one of the most charming pieces of road, finely wooded, and extending through beautiful sylvan scenery to the fine, long stretch of white sandy beach at Barmouth. The day was drawing to a close, 7.30 p.m., when I reached the "Goat Hotel," Beddgelert, embosomed in hills, in a quiet, sequestered, and picturesque spot. I had tea, and a bed made up for me on a sofa, as the house was quite full, being a fashionable place of resort for families and tourists to Snowdon during the season. It is the most handsomely furnished, and altogether the best hotel I have been to throughout my tour in Wales, and within four miles of Snowdon.

Thursday, August 10*th.*—At five a.m., after an early breakfast of bread and milk, I started from Beddgelert, having a fine view of Snowdon as I passed by its base. I went through Carnarvon, Bangor, and over the Menai Bridge; and as I crossed Anglesea in the shades of evening I saw lapwings in the meadows—the first I have met with in Wales—and heard the curlews whistling on the mud-flats as I approached the causeway to Holyhead about a mile in length, I should think; but it was quite dark before I entered the "Royal Hotel," at 9.30 p.m., where I had a bottle of stout, and slept, after my unusually long day's journey of forty-six miles, the longest I had accomplished throughout all my wanderings. On the following morning, the 11th, I embarked from the pier in the Government steamer. We were about six hours on the passage to Kingstown, and from this I went by rail to Dublin. Dined at the "Northumberland Coffee-house," and slept at their adjoining hotel.

Saturday, 12*th.*—Embarked in the *Limerick* steamer.

Monday, 14*th.*—Landed at five a.m. at the Barbican, Plymouth, and on going to the Naval Hospital found no further instructions for me.

During this tour I walked over fifty-eight miles in England, and in Wales 316 miles; making a total of 374 miles. Having reported my return in a letter to the Admiralty, on the 22nd I received a reply from the secretary, stating my appointment had been cancelled.

Saturday, September 30*th.*—I took the coach to London, accompanied by Purser Halse, a former shipmate of mine in the old *Hecla.*

Monday, October 2*nd.*—At two p.m. I attended Professor Cooper's introductory lecture at the London University, and presented Dr. Grant with a specimen of the *Ardea scolopacia* from the island of Cuba; and at three p.m. on the following day I attended his introductory lecture on Comparative Anatomy, and the next day his first lecture on Polygastric Animals. I obtained tickets from Professors Daniell and Rymer Jones for their courses of Chemistry and Comparative Anatomy, which I continued to attend daily throughout the session, combined with a regular course of reading at the library of the College of Surgeons. On the 9th, at two p.m., I attended Professor Daniell's introductory lecture on Chemistry, and on the following day, Professor Rymer Jones's introductory lecture.

November 11th.—I received an appointment to H.M.S. *Thunder*, at Portsmouth, but on the 15th I got it cancelled. My object is to get out in the Antarctic Expedition.

February 8th, 1838.—I delivered a note of introduction I had received from Dr. Robert Willis, the librarian of the College of Surgeons, to Sir Henry Ellis, the librarian of the British Museum, for an admission ticket to the reading-room of the British Museum for six months, and at once commenced my course of reading there with Ackermann's "Tour of the English Lakes," and Dr. James Johnson's "Economy of Health," and spent the evening at the reading-room of the College of Surgeons. On the 14th attended the "Hunterian Oration," at the theatre of the college, by Mr. Travers, from five to six p.m. Sir Anthony Carlisle, president: Sir Astley Cooper and Sir James Clark were present. On the 20th attended Professor Moseley's introductory lecture on Mechanics, at three p.m., at King's College.

March 1st.—My old friend, Lieut. Rogier, R.N., having breakfasted with me, we went afterwards to the Admiralty, where he introduced me to Captain Becher and the Walkers, two brothers, all in the Hydrographical Department. Here we met Captain James Ross, from whom I learnt that the Admiralty had given up the contemplated expedition to the North.

On Good Friday, the 13th of April, I walked to

Woolwich, and dined with my old friend and former shipmate, Captain Charles Gray, of the Royal Marines, at the mess, at six p.m. Colonel Beattie, brother of Sir William Beattie, Lord Nelson's surgeon at Trafalgar, was present.

Tuesday, April 17th.—I attended Professor Daniell's last lecture on Chemistry, which concludes the six months' course, and with which I have been greatly amused and interested.

On the 21st, having called on Captain Ross at his lodgings, I learnt from him that the British Association of Science intended, at their next meeting, to propose an expedition being fitted out for making magnetic observations in a high southern latitude. I again reminded him that I was still a volunteer for it, and my offer was cordially accepted.

Monday, 23rd.—At nine p.m. I attended the meeting of the Geographical Society; Mr. Hamilton in the chair. The Hudson's Bay Company's discoveries formed the subject, and Sir John Barrow rose and said that for twenty years past he had been looked upon as a visionary, but that he still believed in there being an open sea and Polar basin.

Tuesday, 24th.—At four p.m. I attended Stanley's lecture on the "Articulations of the Joints," at the theatre of the college.

May 1st.—At nine a.m. attended Don's lecture (introductory) on Botany at King's College: a heavy and monotonous lecturer.

On the 3rd I was at the distribution of prizes to the medical students at King's College, at two p.m.; the Archbishop of Canterbury in the chair. My old tutor, Sir Astley Cooper, was present, and made a long speech to the students.

May 7th.—I obtained tickets for the courses of lectures on Geology by Phillips, Bell's Zoology, and Taylor's

Forensic Medicine. At noon visited the opening of the Royal Academy in the east wing of the National Gallery. On the 8th attended Moseley's lecture at three p.m., and Professor Owen's at the College of Surgeons at four p.m. On the 10th heard Taylor's introductory lecture on Forensic Medicine and Owen's Comparative Anatomy at the college.

On the 11th, at seven p.m., attended Bell's lecture (introductory) on Zoology, at King's College. On the 14th, at three p.m., Professor Phillips' introductory lecture on Geology. On the 18th, at ten a.m., Partridge's first lecture on Operative Surgery, and Bell's on Zoology in the evening. On Friday, 25th, ten a.m., I went to a meeting at Exeter Hall, on "Negro Emancipation," Lord Brougham in the chair, who spoke for above an hour with his customary eloquence, as did also O'Connell.

Friday, June 8th.—I called upon Sir Astley Cooper, in Conduit Street, who received me with a warm, friendly grasp of the hand. We had a long talk about Polar Expeditions, and he invited me to breakfast with him on any day next week that might suit me, at nine a.m.

Saturday, 16th.—I visited the museum and library of the East India House, lately opened to the public. On the 22nd, about noon, I again called upon my kind-hearted old friend, Sir Astley Cooper, who, upon my saying that it was my intention to go to Scarborough for the benefit of the mineral waters and sea-bathing, said that he had been there lately himself, kindly giving me his card, with the intimation that it might be the means of obtaining for myself a like attention from our professional brethren there he had received.

Thursday, 28th.—I witnessed the coronation procession on its way from Pall Mall to Westminster Abbey, and saw the splendid illuminations in the evening. About the middle of the day a large balloon, with several persons in the car, appeared over Northumberland House, descending so near to the roofs of the houses that sand had to be thrown out, when it rapidly ascended, taking a north-westerly course.

On the 29th I attended the last lecture of Bell's very meagre and superficial course on Zoology. On the day following I attended Professor Owen's last lecture on Comparative Anatomy at four p.m., having to run most of the way from King's College to the College of Surgeons to be in time, as I had been from two till four p.m. at the distribution of prizes in King's College by the Archbishop of Canterbury.

Wednesday, July 4th.—I made my way, through a dense crowd into Westminster Abbey to see the coronation throne, &c., now open to the public.

Saturday, 14th.—The lecture session having now been brought to a close, I got all my tickets to them registered at the Board, Somerset House, yesterday, and having now nothing to detain me longer in town, I started at nine a.m. by the *Duchess of Kent* steamer for Ramsgate, on a visit to my old friend, Lieutenant Rogier, R.N., landing at the pier, between five and six p.m.; and on reaching his residence, No. 16, Royal Terrace, learnt from Mrs. Rogier that he had gone to the pier to meet me, and somehow we missed each other. On his return we had tea, and afterwards took a stroll on the parade. Having passed a week here very pleasantly with my friends, on the 20th I returned by the *Emerald* steamer to town to follow up my summer pedestrian tours.

CHAPTER XII.

In the Lake District—Ascent of Skiddaw and Helvellyn—A rough walk to Ambleside.

July 22nd, 1838.—At one p.m. I started, on the anniversary of my birthday, for the Lake District, leaving town by the Shoreditch Road. My first stay was at Hertford, twenty-four miles, where I slept. From thence my course lay through Cambridge, Oundle, Weldon, Nottingham (visiting Newstead Abbey), Chesterfield, Doncaster, Harrogate, Knaresborough, York, &c. Reached Scarborough on the seventeenth day, Tuesday, the 7th of August, having walked over, in this first stage of my tour, 319 miles. Here I took very comfortable lodgings, at No. 4, Wellington Place, where I remained three weeks, drinking the mineral waters, and bathing in the sea from the open sandy beach of the little bay on the other side of the castle cliff, making excursions from this to Filey, Flamborough Head, Bridlington Quay, &c.

Thursday, August 30th.—I took my departure from Scarborough by Robin Hood's Bay to Whitby; visited the birthplace of our great navigator, Captain James Cook, in the sequestered village of Marton, and Ayton, to which he removed. Passed through Durham to Houghton-le-Spring, where I slept at the "White Lion," and the next morning, Monday, September 3rd, I descended the Belmont coal-mine at Rainton, three miles distant, accompanied by the "viewer," who explained to me everything worth notice, and to whom I had been introduced by a young farmer I accidentally met in the coffee-room of the inn. The depth of the shaft is only fifty-six fathoms, being a new mine, and only been worked for the space of two years, employing about 200 hands with twenty-two horses. The viewer told me that probably 100 chaldrons would be brought up to the pit's mouth to-day, and that when all hands are at work at the seams of coal, averaging four and a half feet in thickness, double the quantity. We groped our way through these burrowings in the carboniferous strata for about three quarters of a mile in a direct line, occupying nearly four hours, it being two p.m. when we again reached the surface. This was to me a most interesting and instructive day's work.

My journey now lay through Sunderland, Shields, Newcastle, &c., to the Roman Wall and military road. At the village of Glenwhelt I took up my quarters for the night at the "Globe," a small country inn, as much farmhouse as tavern, on the roadside. It was without exception the most comfortable and cosy little inn I have ever sojourned at, and the cheapest, notwithstanding the abundant supply of everything, and of the best quality. No charge having been made in the bill for the bed—which was a most comfortable one, with snow-white sheets, in a large airy room, furnished with cases of stuffed birds, maps on the walls, and a fine pair of fossil antlers from the morass below the Roman Wall, ornamented the kitchen mantlepiece—I was told that no charge was made for a bed when refreshments were taken in the coffee-room; and I had an excellent supper overnight, tea, with delicious cream, and with toast and butter, equally good, mutton chops, &c., and the next morning a breakfast in equal perfection, and all for the very small amount of three shillings, the landlord's little daughter waiting on me. The wine and beer-cellar was excavated out of the solid rock. Singular enough, I here saw in the *Newcastle Chronicle* of the 1st the British Association of Science' proposal of a magnetic expedition to the Antarctic Seas. I visited the spa of Gilsland, about a mile distant. This sulphur spring arises in a deep circular glen, enclosed on all sides by densely wooded hills, forming perhaps the most romantic and sequestered spot of the kind in England. I now continued on to Longtown, on the Esk, the last town on the English border, and reached the far-famed Gretna Green in the dusk of the evening. Here I took up my quarters for the night at the "Greta Hall Inn," a most comfortable house. The best front parlour, where the runaway couples have the ceremony performed by the inn-keeper himself, has the model of a line-of-battle ship standing in a corner: the bedroom adjoining communicates by means of folding doors. I had my supper in the back-parlour at the extremity of the hall on the left: a sofa in the room. I slept in No. 3, also at the back of the house, having two windows commanding a fine prospect of the country and hills of Scotland. Beneath the windows quite a farmyard, enlivened by the quacking of the ducks in a pond on the small green. The rooks were cawing in the shrubbery, flanked by the cornfield and stacks. Having slept one night actually on Scotch ground, I again crossed the border on my way to the Lakes through Carlisle. Bassenthwaite was the first lake which met my eye, the peak of Skiddaw appearing over the ridge of hills.

Tuesday, September 11th.—I entered the town of Keswick. Obtained at once a pair of rooms in a cottage at ten shillings and sixpence a week. At seven p.m. had my first view of Lake Derwentwater; and the weather being unfavourable for the ascent of Skiddaw, at eight a.m. on the 13th I made an excursion to Buttermere and Crummock Lakes, going by Lodore and

Borrodale, returning by Newlands; in all thirty-two miles. On the following day, 14th, I walked along the Penrith road to the Druid's Circle, formed of thirty-eight fragments of rocks of various magnitudes, the highest about seven feet. The circle, about thirty-two paces in diameter, having a clump of fifteen fir-trees growing in the centre; the whole enclosed in a green pasture. Journey there and back between three and four miles.

At noon of the 16th I started for an ascent of Skiddaw. Reached the summit, 3022 feet, at 2.30 p.m., where I remained an hour making observations. The shape of the peak from one point of view has much the outline of the deck of a large ship, swelling out at the sides, and as one stands on the imaginary poop at the one end the other tapers off to the resemblance of the bows. A solitary sheep had extended its wanderings even up here, and stood calmly gazing at me on the very summit of the peak. The air was piercingly cold from the clouds of white fleecy vapour sweeping overhead; when all of a sudden, like the rising of the curtain on the stage, there was a break in the surrounding mist, the clouds dispersed, unfolding to the view the whole surface of the country around, producing a very striking and novel effect. On descending I reached my lodgings at 5.30 p.m., having been five hours on the excursion, and gone over some twelve miles.

On the 18th, at ten a.m., I left Keswick for Ambleside, which I reached at 4.45 p.m., passing by several of the lakes—Thirlmere, Grassmere, &c. The finest view of the latter is from Dunmail Raise, a huge gap between the mountains. Saw the poet Wordsworth's residence by Rydal Water; and at about dusk I engaged lodgings, prettily situated about half a mile from the waterhead of Windermere, for a month, at fourteen shillings a week, at "Lowfold Cottage," a white villa, with a good garden and orchard. I had the drawing-room on the first-floor, with the bedroom adjoining. This evening I had my first sight of Windermere; I think the finest of all the lakes, unless it may be Ulleswater.

Having lost some ten days here, from the continued unfavourable state of the weather, I at last started upon an excursion round the Lake of Windermere, going over some thirty miles in twelve hours, leaving Lowfold Cottage on the 29th of September.

Monday, October 1st.—I walked to Kendal and back, a distance of thirty miles in all, by different roads. Had a fine day, returning by moonlight, having been twelve hours on foot.

Wednesday, 3rd.—I made an excursion to Langdale, crossing some fields by a path into the Hawkshead Road from the back of Lowfold Cottage. Had a fine day for the most pleasant excursion I have as yet had, embracing a greater amount and variety of mountain, lake, and valley scenery, with a constant succession of interesting objects; Dungeon Gill Force, flowing between the two pikes of Langdale into an extraordinary chasm, forming a scene of the wildest description imaginable. I returned by the shores of that lovely little lake, Grassmere, as the shades of evening were closing in, having walked over a distance of some eighteen miles.

Thursday, 4th.—Being a fine, clear, bright, sunny day, I made a boating excursion on the waters of Lake Windermere in a small boat lent me by the owner, a neighbouring gentleman. Having cast it off from its moorings in the River Rotha, which runs into the lake, round which I pulled myself in six hours, landing on all the ten islands; commencing with Belle Isle, the largest, consisting of a green sward, beautifully wooded, and the whole isle a perfect little paradise. Reached home at 6.30 p.m.

Saturday, 6th.—Being a beautiful day I took advantage of it to make an excursion to Hawkshead, round

Esthwaite Water to Coniston Waterhead, making the entire circuit of Esthwaite Lake, which is some five miles in circumference. Coniston is six miles in length, by three-quarters of a mile in breadth, and Coniston Old Man majestically towers above all. Returned at six p.m. Walked twenty-one miles.

My ascent of Helvellyn extended to a two days' excursion, returning by Haweswater. I started from my lodgings at eight a.m. on Tuesday, the 9th of October, having come to the determination of attempting the ascent of the mountain from the direction of Ambleside, along the whole ridge to the summit, the customary commencement from Wythburn, eight miles on the Keswick turnpike road, being both the shortest and the easiest ascent. When I had attained to the height of 2950 feet, at Fairfield, the steepest part of the ascent began; and owing to my having all at once become enveloped in one of the dense mountain mists and fogs, I had to steer my way by compass to the summit, which I was four hours in attaining. I was covered with a white hoar-frost, and my fingers benumbed. As I came upon the obelisk here, called "Helvellyn Man," the hoarse cry of the raven (that bird of solitude and mountain heights) was heard, but the bird itself was invisible. As the mist suddenly dispersed for a brief interval, I found myself on the brink of that very narrow ledge called "Striden Edge," rendered memorable from the circumstance of a young tourist having some years ago lost his life in attempting the descent here into Patterdale, being the steepest and most precipitous part of the mountain. Finding the mist again about overshadowing the peak, and no time to be lost, after taking a compass bearing of the Red Tarn beneath me as my guide, I at once began the somewhat perilous descent, whilst I could see my way in the direction of the Tarn, 600 feet below the summit, bounded on the right by Striden Edge, and on the left by Swirrel Edge, terminating in a peak called Catchedecam. On my reaching the valley at the bottom, I came into bright sunshine, whilst the Peak of Helvellyn towered above me, capped in mist, 3070 feet above the sea.

I now followed a mountain stream, passing a desolate-looking village of some half-dozen huts, called Greenside, near some lead-mines, with a few straggling mountain-ash-trees. At the last of the cottages I asked the distance to Patterdale, of two remarkably pretty little rustics, sisters; the elder one really beautiful; and to be the occupants of such a hovel, immured, too, in so desolate a spot! Following the borders of the lake, it was dark before I reached the "Sun Inn," at Pooley Bridge, at 6.15 p.m., where I slept, after an excellent supper in a spacious room, and an equally good bedroom adjoining—altogether a very comfortable inn. I must have gone over fully thirty miles of ground to-day, twelve miles before I left the summit of Helvellyn. Ulleswater is next in size to Windermere, being nine miles long, and one mile wide.

Wednesday, 10th.—At seven a.m. I left the "Sun" at the foot of the lake, by the Penrith Road, passing through Penrith and by King Arthur's Round Table, Lowther Castle, Haweswater Foot, and Kentmere, to Troutbeck and Ambleside. Haweswater is a fine lake, about three miles long by half a mile broad, bounded by lofty mountains. On reaching the "Mardale Inn," about a mile from the head of the lake, I inquired the way to Kentmere, and was informed that there were three different routes to it. One by High Street to the right, now enveloped in thick mist; another by Long Sleddale, by the turnpike road, but some two or three miles further round; and the third, but shortest of all, over Nan Bield. This latter, however, was described as the most laborious and difficult. A pass, lying between High Street on the right, and Harten Fell on the left—in the distance appearing like a mere gap in the summit of a bleak, dismal-looking mountain, over which dark threatening clouds and mist now hung suspended—directly opposite to the inn, and bounding the desolate-looking, swampy, boggy valley, through which a "beck" was winding its course to the lake below. After well-weighing the position in all its bearings, I determined at once on this route. The day was fast drawing to a close—an unfrequented and rugged way before me, with not even the vestige of a track to follow, in the six miles to Kentmere—threatening a dark and stormy, tempestuous night.

At 3.30 p.m. I commenced this toilsome journey from Mardale Green. The weather windy and cloudy, with a lowering aspect, although the rain had ceased. I first struck across the bogs of the desolate-looking valley, in which a few cattle were feeding, and over a stone wall at the base of Nan Bield, from thence began the ascent, which was both steep and rugged, following the course of a mountain stream, which descended along a rocky and swampy surface, getting my feet most thoroughly drenched by the time I reached the summit, from which the wildest scene presented itself, desolate beyond description: a deep and barren peat-clad valley, and a stream struggling through it, enclosed on every side like a wall of steep, bare, frowning mountains, with not even a vestige of bush, tree, or hut to relieve the desert-like aspect, made me doubt whether or not I might have missed my way; and I hesitated, but only for a moment, before I ventured the descent of the almost perpendicular precipice into the wild and dismal-looking hollow beneath me, apparently cut off from the rest of the world. The descent I found even worse than that of Striden Edge, which I accomplished yesterday. I afterwards learnt at Kentmere, that the usual route was round the side of the hill to the left, which, although a more circuitous one, was far less difficult. On reaching the bottom, often sinking to my knees, I struck across the valley for a ridge, which I had hoped promised a better state of things; but on gaining the summit of this, a second edition of a similar character presented itself. However, persevering onwards, I at last came in sight of a pile of slate, indicative of the vicinity of some mining operations, near a cavern-like opening on the opposite mountain-side. As the night was fast closing in, I congratulated myself on the chance of at least obtaining a shelter from the elements within its recess, should I be benighted in this mountain-girt hollow, unable to find my way out. Still, never relaxing my exertions, I followed a cart-track from the supposed mines, much wearied, when, on turning an angle of the rocks, I fell in with a man with a horse and cart, from whom I learnt that I was just on the confines of Kentmere; and following the cart-track for about half a mile it led me into a lane with trees and bushes, and a small farmhouse: a little beyond this a hut. Here I inquired my way of an old man, from whom I could elicit but scant information, till a young woman, coming out of one of the outhouses, pointed out to me the direction in which Troutbeck lay—over another dark frowning mountain pass, almost hidden in clouds and mist. It was now between five and six p.m., and the shades of evening fast setting in. She advised me by no means to attempt the crossing of this mountain pass to-night, as the way was difficult, and it would be dark before I could reach Troutbeck. I next proceeded across some fields and a meadow, over hedge and over ditch, and across a mountain rill, to the road at the base of a hill, where I passed the last cottage of Kentmere; and was directed to continue along the cart-track, up the hill at the back, from which I could see the summit of the mountain, wrapped in clouds and mist, which I had yet to pass over.

It now wanted only a quarter to six o'clock, with four miles yet before me to Troutbeck. With the increasing dusk the rain began to fall fast, and continued so to do over the summit the whole way, as I followed the cart-track. The night had now set in very dark and tempestuous, but the worst part of my journey was already surmounted, the road winding between two stone walls to the "Flow," half a mile from Troutbeck, along a considerable descent, passing through two or three gates—still nothing but the wall on either side, and hills frowning above, with mountains in the distance, to which I almost began to think there would be no end. About two miles before I reached Troutbeck the lights of the village appeared, studding the opposite mountain-side, on the right, with what appeared to me, through the gloom of night, a valley between. The night, although still stormy and windy, began to clear, and the stars made their appearance in the intervals between the densely black clouds, which drifted rapidly across the blue sky. On passing through another and the last gate, the walls were succeeded by a railing, when I entered a wooded lane, with a light in front of me, which proved to be the "Flow," and the residence of a captain in the navy, on the left side of the road, half a mile from Troutbeck. It was now 6.30 p.m.

Passing down a dark, shady walk into the road, I passed by the straggling village of Troutbeck, and on reaching the finger-post took the road to the right, which two miles further on brought me into the Kendal and Ambleside Road, just above "Lowwood Inn." Passing by the lake, I reached Lowfold Cottage at eight p.m., having been thirteen hours on foot since I left the "Sun Inn," at Pooley Bridge, this morning at seven o'clock; and right glad was I to find myself once more in my snug little room by a fire, seated at table with a soothing cup of tea, and hot mutton-chop, and an equally comfortable bed to turn into; which in the earlier part of the evening I had almost begun to despair of. The distance

I had walked over to-day amounted to about thirty-eight miles, which, with some twenty-six miles yesterday, made a total of sixty-four miles.

Friday, 12th.—Being a fine, clear, windy day, I went to Kendal to receive my quarterly half-pay, by a bill of exchange. Walked the fourteen miles there in three hours, making twenty-eight miles there and back.

Saturday, 13th.—I visited Stock-Gill Force, a waterfall 150 feet in height, having its rise in the Ambleside Hills; falling in two cascades, half a mile from "Salutation Hotel."

Wednesday, October 17th, 1838.—I took my leave of Lowfold Cottage after breakfast, amid showers, threatening a stormy day, and brought my first day's journey to a close at the Flockborough "King's Arms Inn," where I slept, having passed by Coniston Water, Ulverston, and across the Leven Sands, after a journey of twenty-eight miles. On the following day I only got through nineteen miles, owing to stormy weather. Blowing a gale of wind, and the rain falling in torrents nearly the whole of the way to Lancaster, where I slept at the "Royal Oak," in the market-place. On changing my wet clothes I found my feet greatly swollen, accompanied by a violent headache in the evening. Altogether I have rarely experienced a more trying and miserable day's journey than this has been; but was fortunate in finding so comfortable an inn. I had to wade over the Lancaster Sands.

Continued my journey through Preston, Bolton, Manchester, and other towns to Derbyshire; passing through Castleton and the wild and picturesque romantic pass of the Windgates, a precipitous cleft in the mountain-side, and the ruins of Peveril Castle, the scene of Sir Walter Scott's charming tale, "Peveril of the Peak." From this my course lay by Buxton, Bakewell, Chatsworth, Matlock, and Ashbourne. From the latter place I made an excursion to Dovedale, along the banks of the river Dove, across which three kingfishers flitted like emeralds, as their glossy green plumage reflected the sun's rays. At the village of Thorpe I had a lunch of bread and fresh butter, with a pint of home-brewed ale, handed me by the landlady's little daughter, the prettiest girl, with her raven locks, I have seen throughout my tour.

Passed through Derby, Leicester, Northampton, St. Alban's, Barnet, and Highgate, to London. Darkness overtook me in passing over Highgate Hill, Camden Town, Tottenham Court Road, to St. Martin's Lane, to my old lodgings, No. 2, Northumberland Court, which I reached at eight p.m. On my sixteenth and last day's journey I had walked forty-four miles. The total number of miles walked over in my tour were :—

From London to Ambleside . .	557
,, Ambleside to London . .	337
Branch excursions . .	306
	1200

My journey from Ambleside was performed in about a fortnight, having returned on Thursday, the 1st of November; and the whole tour occupied a little over three months. And now, in the aggregate, I find that during the preceding four years my pedestrian wanderings have extended over some 3440 miles, all on foot.

CHAPTER XIII.

Appointed to the *Erebus* for the Antarctic Expedition—The Queen's *Levée*—Final arrangements—Departure.

Monday, March 11th, 1839.—The Antarctic Expedition may now be considered fairly decided upon. Having this evening attended the Geographical Society's meeting, Mr. Greenough in the chair, a portrait of Weddell, a half-pay master in the navy, who had attained the highest southern latitude some years ago, was presented to the Society by one of its most distinguished fellows, Mr. John Brown, accompanied by a letter urging the Society to lend its aid in sending out an expedition, when Sir Roderick Murchison rose and said he believed the Government had settled that an expedition should proceed immediately to the Antarctic Seas, under the command of Captain James Clark Ross, through the application of the Marquis of Northampton.

On Wednesday, the 10th of April, at eight p.m., I received my appointment to H.M.S. *Terror*, dated the same day; and on the 15th I went down to Chatham to join her. Took up my quarters at the "Sun Tavern," and on the following day took up my commission at the dockyard, and reported myself to Captain Clavell, the superintendent of the dockyard. Went on board the *Terror* and *Erebus*, both in dock. On the 23rd met Captain Ross at the "Sun," when he told me I was to have charge of the zoological department, and Joseph

Hooker, the assistant-surgeon, a son of Sir William Hooker, to have the botanical. We were hulked on board the *Renard*, an old packet brig. On the 25th, meeting Captain Ross in the dockyard, I took the opportunity of calling his attention to the geology.

Wednesday, May 1st.—I went to Lucas's naval and military outfitting warehouse, and ordered a glass, &c. A fine, bright, sunny, warm day. On Saturday, the 11th, the *Erebus* being ready for hoisting the pennant, we were turned over from the *Terror* to her, and Captain Crozier with his officers to the latter ship; and I took up my appointment to the *Erebus* at the dockyard. Saw our old North Polar boats in the boat-house.

On the 17th we shifted over to the *Tartar* hulk lying off the dockyard. On the 21st the first lieutenant showed me a memorandum from Captain Ross to the secretary of the Admiralty, in reference to the appointment of a naturalist, in which he stated that he did not consider the appointment of a special naturalist to the expedition at all necessary, as the surgeon of the *Erebus* had devoted his time and attention for many years past both to the study of zoology and geology, and specially qualifying himself for the position, whilst Dr. Hooker was only known as a botanist; and that a draughtsman specially appointed would be far more desirable;—and this day Dr. Hooker joined us as assistant-surgeon and botanist. On the 27th, Captain Crozier, dining with us, told me that the application for the appointment of a special naturalist was smashed; nor should I here have thought it worth while alluding in any way to these bygone events, which in no way affected the friendly feeling existing between Dr. Hooker and myself throughout the long and somewhat trying voyage we were together, had not I felt, on our return to England, that the same influence stood in the way of my being permitted any share in bringing out the publication of the collections of natural history which I had made with so much toil and labour, not altogether free of personal expense to myself, to say nothing of the risks frequently incurred.

On the 30th I was launched in the *Fantome*, a beautiful new sixteen-gun brig. There were a great number of ladies present, and amongst them the belle of the place, the daughter of our outfitter, the pretty young Jewess, Annie Lucas. On the following day I went to town.

June 1st.—Called on Sir Astley Cooper, and on Professor Owen.

Wednesday, 26*th.*—I attended the Queen's *levée*, for which I had purchased a new sword, knot, belt, and cocked hat, and obtained an introduction from Sir John Pechell. It was a showery day; the *levée* hour two p.m.; I was half an hour late. Notwithstanding, I got in with the first batch, and it being my first *levée* I knelt and kissed her Majesty's hand. Left about three p.m.: a dense crowd. On returning to my lodgings to sleep, I found a packet of letters awaiting me from Captain Ross. containing introductions to Professor Owen, Mr. Lonsdale, of the Geological Society, and an invitation from Dr. Fitton, the geologist, to spend two days with him at his house in Norwood; for which I started on Saturday, the 29th. In looking over the old folio edition of "Cook's Voyages," Dr. Fitton more especially called my attention to Kerguelen's Land, an island he appeared to take much interest in, and wished me to very carefully explore, presenting me with his old and favourite geological hammer, cutting an inscription on it with his knife, and gave me also a small volume of his geological survey of Hastings, during which he had used this hammer in his examination of the rocks.

After my visit, on the 2nd of July, I attended the Geological Council of the Royal Society, at two p.m., with Captain Ross, Dr. Fitton (president), Dr. Buckland, Mr. Lonsdale, and other fellows present.

Several papers were read. Dr. Buckland gave me his card, and a note of introduction to Mr. Fox Talbot, the inventor of the photogenic drawings, &c. The meeting broke up about 3.30 p.m. On the following day I called on Mr. Talbot, at No. 31, Sackville Street, and had about an hour's conversation with him respecting his new process, of which he presented me with a specimen, and an invitation to spend a day or two with him at his country house, where he promised to give me a practical lesson on the whole process. He also showed me a representation of his house, which occupied him for four hours in finishing.

Friday, August 32rd.—Dr. Andrew Smith, from Fort Pitt, came on board to see me; and whilst showing him round the ship, Captain Ross, coming on board at the time, I introduced them to each other. On the following morning I breakfasted with Dr. Smith, and he showed me round Fort Pitt, its library and museum.

September 2nd.—Earl Minto and the Lords of the Admiralty came on board and mustered both ship's companies. On the 13th, when in town, I met Sir Edward Parry in Regent Street, when he took my arm, and walked as far as Argyle Place with me, offering me his hearty congratulations and wishes for the success of the voyage I was about making.

Thursday, 19th.—At nine a.m. we cast off from the hulk and made sail down the river amidst heavy rain, and secured the ship opposite the church of Gillingham, near to the *Terror*.

September 24th.—Cables on board, and to sail tomorrow, so I went on shore to take my last leave of Chatham, and at midnight went on board from Gillingham Reach. On the following day made sail down the river and anchored off the *Mouse* near the Nore, for the night. Captain Ross's father and mother having come down the river with us, I asked him and them to join us at dinner in the gun-room at 3.30 p.m., where they would meet Beverley, our old Arctic shipmate, who had been my guest down the river. Thirteen of us sat down to dinner in our small mess-room, myself taking the head of the table. Just after dusk the *Terror* came to an anchor astern of us. The next day we again got under weigh and anchored in Margate Roads, from which I could see very plainly from the deck my old station, Epple Bay Watch-house. We were detained here by westerly winds until Monday, 30th, when the wind shifting round to the eastward, we took our final departure at seven p.m.

CHAPTER XIV.

On half-pay again, 1843—Applications for promotion—Fellow of Royal College of Surgeons—Appointed to the yacht *William and Mary*.

Saturday, September 23rd, 1843.—The *Erebus*, after an absence of four years from England (my journal for which period is recorded elsewhere), was paid off, and I left her hulk, the *Salsette*, for town, and went to my old lodgings at No. 2, Northumberland Court. And on the 26th of October I left town by coach for Plymouth, and found my mother and sisters had during my absence removed from Stoke to Densham Terrace, Plymouth.

December 7th.—I sent in a memorial to the Admiralty, through Sir Edward Parry, asking for my promotion to the deputy-inspectorship; and on the 13th received an answer that their lordships could not comply with my request.

As the metropolis was now likely to become my home for some time to come, I went to Dover on the 20th to look for a suitable cottage for my mother and sisters, so as to have them nearer to me than the West of England, and yet on the sea-shore, to which they had been all their lives accustomed. On the following day I left Dover for Ramsgate; dined with my old friend there, Lieutenant Rogier, and the next morning went on to Margate. Embarked there in the *Red Rover* for London. On the 23rd I went by train to Reading, and returned to town the same evening.

January 29th, 1844.—A vacancy having taken place in the *Royal Sovereign* yacht at Pembroke, I applied for it, but was unsuccessful, though amongst the numerous candidates I stood second for it. The fortunate candidate obtained the appointment entirely through being my senior on the list.

March 17th.—I embarked in the *Sirius* steamer for Plymouth, anchoring the following day in the Sound after dark, and I did not reach Densham Terrace till 10.30 p.m. Found that my poor old spaniel Hecla, whom we had had for some seventeen years, and who was very infirm from age when I was last here, died at one p.m. yesterday. On Thursday, 21st, I enclosed the faithful old dog's remains in a half-barrel, and inscribing on the lid his name, age, and date of his death, buried him in the garden at the back of the house, left corner.

On the 12th of June I met Sir John Franklin at the Geological Society's meeting, and spoke to him, congratulating him on his return from Tasmania.

January 17th, 1845.—I met my old friend and brother officer, Sir John Hammett, at the Admiralty, where he had been seeing Lord Haddington, to thank him for his promotion to the rank of deputy-inspector general. I afterwards called on Captain Ross at Elliott Place, Blackheath, where he introduced me to his wife, having married since the return of the expedition. Old Wall, our old cook in the *Erebus*, opened the door to me.

On the 24th, Captain Hamilton succeeded Sir John Barrow as secretary to the Admiralty.

Saturday, February 1st.—When at the college library I first became introduced to my late lamented friend Dr. Daniell, who had distinguished himself as an African traveller.

March 18th.—I met Captain Crozier in the Strand, from whom I learnt that Fawcett, one of my boat's crew at Kerguelen's Land, and old Wall of the *Erebus*, were both going out with Sir John Franklin.

On April 7th I had my first interview at the Admiralty with Lord Haddington. He received me very kindly, and listened attentively to my somewhat lengthy statement of my services; but, at the same time, intimating to me the difficulty in the way of promotion in my class, from the promotion to the deputy-inspectorship being considered as great a step as the promotion of a captain to flag rank, in a list kept so small and select.

April 22nd.—Dr. King, who has distinguished himself in Arctic discovery and in the search for Franklin since, was introduced to me by Dr. Daniell, at the office of the Statistical Society, in Waterloo Place; both now numbered with the dead. On the following day, at eight p.m., I accompanied Dr. Daniell to the Ethnological Society's meeting, in Sackville Street, where Dr. Richard King, as the honorary secretary, read a paper to the Society, and Dr. Wolffe, the distinguished traveller to Bokhara, gave a long account of his travels. Dr. Hodgkin, the Quaker physician and geographer, was also present, to whom Dr. Daniell and myself were both introduced by Dr. King. Admiral Sir Charles Malcolm was in the chair.

July 7th.—Calling on Dr. Lindsay, at No. 3, Trinity Court, he showed me a recent Admiralty regulation, requiring of all candidates for the inspectorial ranks a certificate of having gone through a course of dissections and surgical operations, subsequent to the date before they could be promoted, and he, himself, although already a deputy-inspector, was complying with the order, at the Westminster Hospital, under Professor Hunter, to whom he introduced me; and so soon as a subject was obtained I followed his example; when, by unremitting exertions in the dissecting-room, in less than a week I obtained my certificate, accompanied by a very flattering letter, which I at once transmitted to the Admiralty in time to join the yacht.

I received my appointment to the *William and Mary* yacht on September 1st, and was told that there were no less than fifty candidates for her.

September 8th.—I went to Woolwich, and took up my commission in the dockyard. Met the surgeon, Mr. John Edwards, who had also been one of Parry's followers, and whom I now superseded in the yacht, and we held a survey on the medical stores, when I learnt from him that 1*s.* 6*d.* a day was allowed in lieu of quarters, and I at once engaged lodgings at No. 2, Woodland Terrace, and Edwards went to No. 11, Craven Street, which I had recommended to him.

October 3rd.—I reported myself to the superintendent of the dockyard, Sir Francis Collier, he having been absent on leave when I first joined.

Tuesday, 21st.—Being the anniversary of the battle of Trafalgar, Dr. Suther, the surgeon of the dockyard, as one of the survivors, gave a dinner on the occasion, at which I was present, and sat next to another old brother officer, who had been in the *Royal Sovereign* in the action, Dr. John Clark, and at the end of the war had his name struck off the list for refusing to serve when called upon; and had been in practice at Weldon, in Northamptonshire, in the same house, for the last thirty years. He took from his pocket and showed me his appointment to the *Royal Sovereign*, dated October 14th, a week before the battle, and bearing Nelson's signature.

Wednesday, March 11th, 1846.—At two p.m. I attended Lord Haddington's *levée* at the Admiralty. He received me very graciously, advancing from his seat to the middle of the room to meet me, shaking hands, and pointing to a seat opposite his own, requesting me to be seated. He asked me many questions about the Antarctic regions, and on my asking him for my promotion to the deputy-inspectorship, he appeared to be quite unacquainted with the nature of the new inspectorial ranks and the duties of the yacht, all of which I had to explain to him. His manner was both attentive and courteous, ending by assuring me that I might rely upon his bearing in mind all I had said. The interview lasted about ten minutes.

On the 27th I went to Gillott's, my tailor's, and brought down with me the new uniform : all classes now to wear two epaulettes.

July 18th.—Captain Houston Stewart, the newly-appointed superintendent of the yard, joined. I saw him in his office about noon, when he shook hands with me, and asked me to be seated.

September 1st.—I left by train for Chichester, and from thence to Portsmouth by the coach, and slept at the "Star and Garter," Point. On the 3rd, after having had a bathe in the sea, I left the Victoria Pier by a steamer, on an excursion round the Isle of Wight; had delightful weather, just eight hours in the circumnavigation of the island, returning to Victoria Pier at five p.m. Bonchurch and Shanklin appeared to be the most beautiful spots. On the 5th I crossed over to Ryde and made a tour of the island, making the "Marine Hotel," Ventnor, my headquarters. I rambled about the interior of the island, by Appuldecombe, Arreton, Newport, &c., returning from Cowes to Portsmouth on the 11th, and slept in my old room, No. 30, at the "Star and Garter." On the following day I sought an interview with Commander Crispin, of the royal yacht, *Victoria and Albert*—who had been a shipmate of mine in the *Icarus*, on her passage home to England from the West Indies, many years ago—to ask him about a vacancy expected soon to take place in the Queen's yacht. I met him coming out of the commissioner's office in the dockyard; unfortunately he was pressed for time, stating that a signal was made for him, and in the hurry I could only exchange a few words with him, when he told me no rumour had reached him of any contemplated change, and that great interest would be made for the appointment of surgeon of the yacht; nor was he aware whether Lord Adolphus Fitzclarence had any one in view himself. Disappointed in getting any information with reference to an appointment I was anxious of becoming a candidate for, I returned in the afternoon by train from Gosport to town, and slept at the "Ship Hotel," Charing Cross; returning to Woolwich on the following day, when I found a letter dated October 5th, "Globe Hotel," Portsmouth, on my supper tray, from my young friend Gregson, bidding me adieu on his return to Tasmania in the *Windermere.*

On Wednesday, the 7th of September, received a communication from Sir James Ross, asking me to contribute a Geological Appendix to his narrative of the Antarctic Voyage, which I answered at once. Meeting Captain Houston Stewart, coming out of his office, who very cordially shook me by the hand, I stated to him that I was desirous of commencing my leave of absence to-morrow, when he said he also intended going on leave the same day. Having various things in town to attend to, I passed my fortnight's leave there. On my return to my duties, on the 23rd, I wrote to Sir James Ross, and on the 29th I received a packet from him containing my notes on the geology of Tasmania, with a note, and the following day I began correcting them for the press. Finished them on the 4th of November, having employed my evenings after the duties of the dockyard were over about them—often till midnight. I waited on Sir Francis Collier, who had returned to his dockyard duties, last night.

November 14th.—The yacht this morning sported the blue flag at the mizzen, Sir Francis having hoisted his flag at daylight, on his promotion to the rank of rear-admiral. I went on board the yacht. Saw the appointment of Sir Edward Parry as superintendent of Haslar Hospital, and Sir Gordon Bremer is to succeed Sir Francis Collier here; Captain Houston Stewart having been appointed controller of the coastguard.

December 4th.—At work about my Antarctic Geological Sketch, which I finished on the evening of the 11th.

Her Majesty's Yacht *William and Mary* at Woolwich.

CHAPTER XV.

Transferred to the *Fisguard* frigate—Unfair treatment by the Director-general—Sir Gordon Bremer—Am superseded.

On the 29th of March, 1847, wrote to Sir James Ross, and returned him all the revised sheets for the appendix of his book.

May 19th.—Meeting Sir Gordon Bremer on board the river steamer as I was going up to town to-day, he having got into conversation with me, I took the opportunity to mention to him that it was my intention to offer my services to the Admiralty to go in search of Sir John Franklin, when he kindly offered to forward my application.

Monday, June 21st.—I received a presentation copy of our Antarctic Voyage from Sir James Ross, and on Monday, the 9th of August, I went to the Admiralty and received back my private journals and sketches of the Antarctic Expedition from Captain Miles, who had charge of the box containing them and others, and which had only just arrived from the "Ship Hotel," where they had been lying ever since the return of the expedition. The box had not been opened, and he expressly unpacked it to give me mine.

Friday, August 13th.—Dr. Fitton, with his sons and daughters, and a party of his friends, came down to Woolwich at noon. The whole party, consisting of Dr. Fitton, his sons and daughters, Sir James Fellows and his daughter, Professor Studer, a Swiss geologist, &c., lunched in the little room at my lodgings. After we had lunched, between two and three p.m., we went to the dockyard about 3.30 p.m., and having made the tour of it, took the young ladies on board the *Royal Albert*, of 120 guns, building in the shed. They being very anxious to explore the interior of so large a ship, a great novelty to them, I handed the two eldest Miss Fittons, and their cousin, Miss Fellows, down the hold; no small effort for them in a ship whose tonnage was 3393 tons. We passed near the yacht, visiting the new surgery and new dock, finishing with the boathouse, containing Sir John Ross's old boat; and at 5.45 p.m. saw them all off from Charlton Pier by the steamboat, my sister and myself returning home to tea; and on the 20th I saw her off from the London Bridge terminus to Dover.

October 19th.—On going to the office this morning I found that I was transferred from the Yacht to the *Fisguard* frigate. I hesitated about taking up this appointment, for it would place me in the ordinary routine of a three years' appointment, and I had held the yacht as a life appointment; at least it had hitherto been always so considered, and I was distinctly told in the office when I first joined her that it was for life. I therefore memorialized the Admiralty for promotion to the deputy-inspectorship, on the yacht being paid off, in compensation, and on my consulting Sir Gordon Bremer about it he offered to present my application, assuring me that he had considered my promotion so certain that he had already promised the vacancy occasioned by my promotion to Captain Duntze, the late captain of the *Fisguard*, for his surgeon, Dr. T. Russell Dunn. He advised me to go at once to the Admiralty and see Admiral Prescott, the medical lord, about it; which I did, and strongly remonstrated against such treatment of just claims. But all the satisfaction I could obtain was that it was Lord Auckland's act, and that he, i.e. Admiral Prescott, had no power or influence in the matter. Sir Gordon Bremer now gave me a note to Captain Eden, the first lord's private secretary, in order to obtain for me an interview with Lord Auckland himself. On presenting this I was told by Captain Eden that Thursday was the only day on which his lordship saw officers, adding that the director-general denied the permanency of the appointment, and further that I was only turned over to the *Fisguard* to complete the remainder of the three years not completed in the yacht. So that it appears Lord Auckland leaves everything in the hands of the director-general, and in such hands as his the grossest injustice is perpetrated and perpetuated.

On Thursday, the 28th, by putting on great pressure, I managed to obtain an interview with his lordship, useless, as I foresaw it would prove, prejudiced as he was by the head of our ill-managed department. On my entering the room his lordship coldly pointed to the chair opposite to him, to be seated, when I stated in as brief a way as possible my present position; that I had held the yacht as a life appointment, consequently my transfer to the frigate to complete time was a loss to me of my position in the service, and the only other equivalent appointment was the Queen's yacht; therefore I had been led to hope that with my special services, and peculiar position, promotion to inspectorial rank would have been conferred upon me on my ceasing to hold a life appointment; adding that I had some months ago volunteered to conduct a party to clear up Sir John Franklin's fate, and was still as ready as ever to go. But I never before had an interview with a more reserved, colder, or more repulsive official. His lordship went on with the writing before him, without even raising his eyes from the paper once whilst I was addressing him, and on my rising to go, his only reply was that *he* did not consider I had lost by the change.

The good old commodore having failed in his well-disposed efforts to serve me, now strongly advised me to take up, as a final alternative, the appointment, rather than by futile opposition risk injuring my future claims on the service and prospects. All my friends being of the same opinion, the prudence of which was but too obvious to myself, I at last took up my commission, lying at the dockyard office.

I now resolved to make every effort in my power to get charge of an expedition in search of Sir John Franklin and my old ships, the *Erebus* and *Terror*. Accordingly, on the 1st of December I made a second offer of my services in a letter, volunteering for any exploring expeditions that might be sent out; taking my letter to Sir Gordon Bremer for his approval, and to forward to the Admiralty for me.

Dr. Richard King having got up some public meetings in furtherance of the search for Franklin, I attended one at the Western Institute, Leicester Square, at eight p.m., at which he lectured to an audience of about 300 ladies and gentlemen.

January 4th, 1848.—I accompanied Dr. King to the Russell Institution, where, at eight p.m., he gave another

Arctic lecture; and again, on the 27th of April, at Crosby Hall, in Bishopsgate Street.

Thursday, May 28th.—Between nine and ten p.m. I went to a ball at Dr. Fitton's, in Harley Street, to which about 100 guests were invited. Dr. Fitton there introduced me to the great mathematician, Dr. Babbage. I left at 1.15 a.m. and walked the whole of the way home to Woolwich, some eleven miles. A beautiful night, or rather morning, with a clear blue sky. I heard the cuckoo, the lark, and the nightingale singing, reaching my lodgings at 3.45 a.m.

August 22nd.—Lady Bremer and her friend Miss Webber, the bishop's daughter, came on board the *Fisguard*, to whom I was introduced by Sir Gordon.

At her ladyship's ball, supper was laid on a table on the main deck, decorated with transparent lanterns and laurels; the *Fisguard* being entered by means of a floating-bridge from the dockyard. There were from seventy to eighty guests, and I left about eleven p.m.

September 2nd.—Whilst on board the *Fisguard*, I was sent for by Sir Gordon, for the purpose of consulting me about the complaint (diabetes) he was labouring under. Having suffered much for the last six weeks he was getting alarmed, and wished me to prescribe a course of medicine and diet for him, at the same time requesting me to write him a note, recommending a change of air, which he could show to Admiral Dundas on asking for leave of absence. On my suggesting to him that it would be a breach of etiquette on my part to do so, when the surgeon of the dockyard was, by the rules of the service, his legitimate medical attendant, he replied that was his affair, and he felt himself at liberty to place himself in the hands of any one in whom he had the most confidence; therefore, considering it in the light of an order, I wrote him out the desired note.

On Sunday, the 10th, after divisions on board, the

commodore said he wished to see me in his cabin on the subject of his health. On entering he requested me to take a seat, and then went on to say that he had already in the last few days felt himself better from the change I had made in his diet, together with the course of medicine I had placed him under; at the same time telling me he was about paying a visit to Tunbridge Wells, and I then strongly recommended him to avail himself of so favourable an opportunity for taking the chalybeate waters. On the 12th Sir Gordon invited me to dine with him in the evening. On the 15th, calling at the commodore's house to see him, I met Lady Bremer, from whom I learnt that Sir Gordon had obtained six weeks' leave. She asked me into the drawing-room, and had a long conversion with me as to his health. On my return on board afterwards, I saw him, when he expressed himself very gratified with the treatment under which I had placed him, and said that it had met with the entire approbation of Dr. Drummond, of the Royal Marine Hospital, who had seen him as an old friend. On Sunday, the 17th, I saw the commodore at his house; Lady Bremer and her friend, Miss Webber, were in the library with him. Sir Gordon, requesting me to take a seat by the fire, thanked me warmly for my attention to his case, and, shaking hands, said he left to-morrow; and Lady Bremer promised to write to me whilst away and report progress; and on Monday, October 9th, I received a communication from her ladyship and Sir Gordon, giving a very satisfactory account of his progress and the benefit he had derived from the mineral waters.

November 3rd.—Sir Gordon having returned from leave last night, I saw him to-day on the subject of his health and retirement from his public duties, as affording him the best chance of recovering his health. The Tunbridge waters he spoke of in the highest terms, and thanking me in a very complimentary manner. On the

9th I met him in the dockyard after his interview with Dr. Bright, which I had suggested to him before he left, when he told me that the doctor advised him to continue the course of treatment from which he had derived so much benefit, and that he did not know anything further he could suggest to him, unless it was to add bottled stout to his diet.

Friday, 17th.—Yesterday the commodore was superseded by Captain Eden, and he will leave Woolwich to-morrow. I called at his house to take leave of him. Requesting me to take a chair by the fire he entered into some details in reference to his state of health, again warmly thanking me for my attention, and cordially shaking me by the hand and accompanying me across the yard to the office steps. We parted, as the sequel will show, for the last time.

December 25th, Christmas Day.—I dined with my friend Fowler, who was flag-lieutenant to the late commodore, Sir Gordon Bremer, and at his table met an old shipmate of his, Commander Hall, now Admiral Sir William King Hall, and spent a very pleasant Christmas evening, leaving at eleven p.m.

Sunday, 31st.—Having been superseded, I attended divisions on board for the last time to-day in cocked hat and sword, and gave up charge to my successor. Commodore Eden, the new superintendent, mustered the ship's company.

CHAPTER XVI.

The Auckland Whaling Company—My plans for a Franklin Search expedition—Lady Franklin's approval—A holiday on the continent.

Thursday, January 4th, 1849.—Having received an introduction to the Enderbys, ship-owners, of No. 13, Great St. Helen's, Bishopsgate Street, I called on them, and met with a cordial reception. Mr. Charles Enderby, who is going out to the Auckland Islands to establish a fishery there, had a long conversation with me on these islands, asking me if I had any drawings of them, which I promised to furnish him with. He gave me an invitation to dinner to meet some friends interested in the subject.

On the 10th, after my return from town late in the evening, I found Mr. George Enderby awaiting my return at my lodgings, to ask me to come on Wednesday next to their dinner, and at six p.m. of the 17th eight of us sat down, including the two brothers, at the octagon table in an octagon-shaped room. I sat next to Colonel Colquhoun, having Professor Airy, the Astronomer-Royal, opposite to me. Mr. Charles Enderby showed us a New Zealand Tui, or parson-bird, in a glass case, which he had kept alive in England for two years. He also showed us a bedstead of King Henry the Eighth and Anne of Cleves in a fine state of preservation, bearing the date of 1539 on an inscription at its head, curiously carved and inlaid throughout in the Old English style. It had been for a century in the possession of

Enderby's family, having belonged to his grandfather. I left at ten p.m. in company with Professor Airy, whose way home lay in the same direction as my own. The night was fine and starlight. We parted company on the Greenwich Road, I taking the lower road to Woolwich, reaching my lodgings at 10.45 p.m.

February 14th.—I took my sketch of the Auckland Islands, which I had made for Charles Enderby, to their office Great St. Helen's. He seemed much pleased with it, and said it gave him the best idea of the islands he had yet seen, and that he should get it engraved for his account of the islands. On the 22nd, about eight p.m., when in Peckham, and returning over Blackheath to Woolwich, I saw the finest display of the aurora I ever witnessed in this country. The night was bright starlight, with neither moon nor cloud to lessen the effect. A deep crimson blush overspread the horizon to an altitude of about 45° westward of the "Great Bear;" from this, flickering tapering rays of a beautiful pale primrose radiated upwards, changing their position every moment. This brilliant display lasted for about ten minutes, leaving for some time afterwards a pink hue in that portion of the horizon.

On Saturday, the 24th, I removed from Woodland Terrace, to my old lodgings, No. 11, Craven Street, Strand.

On Tuesday, the 27th, I attended the *levée* of the First Lord of the Admiralty. My name was called about 2.20 p.m., and I remained about six minutes. Sir Francis Baring is a stout, portly, jolly-looking, red-faced man, having an agreeable, mild expression of countenance. He rose from his seat on my entrance, shook hands, and requested me to be seated in an armchair, opposite to him. Having at once told him that my object was promotion, briefly stating my claims in the Polar Expeditions, and in which I had saved the Government the expense of a naturalist and geologist, and had volunteered to go in search of Sir John Franklin, in command of a party. He replied that only a provision-ship would be sent out, and asked me if I did not think once going out enough. I told him that I was the only officer left unpromoted in the Antarctic Expedition; and rising to go, he again shook hands, and asked me for my card, which he said he would keep.

On Friday, March the 23rd, at three p.m., I called at Enderby's office, and accompanied Charles Enderby to the Auckland Whaling Companies' Office, in Cornhill, where I was introduced to several gentlemen associated in the undertaking, and underwent a long examination, before the president and directors, as to the various capabilities of the Auckland Islands for a settlement; such as the character of the vegetation, soil, rocks, grasses, building materials, rise and fall of tide, depth of water, best parts for grazing cattle and sheep, and the number of wild pigs already there. Having replied to all these questions, I pointed out the north side of Laurie Harbour as the most promising site for a settlement. I was seated at the table on Enderby's left, with the president, an old gentleman whom I had met at Charles Enderby's dinner-table, opposite to me, in the centre of the other side of the table. I left at four p.m.

On the evening of the 26th, I paid Professor Queckett a visit at No. 15, Dorchester Place, Dorchester Square, where I met Charles Enderby. The whole evening was occupied with the microscope, and Enderby and myself left together.

Monday, April 2nd.—At two p.m. I called on Lady Franklin, at No. 4, Spring Gardens, to offer her whatever assistance I was able in the search for her lost husband. Her niece, Miss Cracroft, and Mrs. Leaves—Sir Edward Sabine's mother-in-law—were with her ladyship in the drawing-room. Lady Franklin said she had not forgotten me, although some eight years had elapsed since we last met in Tasmania. She entered into a long conversation about the various plans of search for the lost ships, for nearly two hours. She told me that when she was at the Admiralty she had named me as well suited for the charge of any searching-party that might be sent out; and seemed very desirous that the Wellington Channel should be thoroughly searched, as Sir John Franklin had before his departure said he would attempt that channel. She also wished me to see Admiral Beaufort, the hydrographer, on the subject of my plan for the exploration of Jones' Sound.

Monday, 9th.—In compliance with Lady Franklin's earnest wishes, I had an interview with Sir Edward Sabine, at his residence on Shooter's Hill; I showed him my plan of search, which he read, and highly approved of. Lunch was placed upon the table for me, in which Lady Sabine's mother joined me. On leaving, Sir Edward presented me with a copy of Lady Sabine's translation of Humboldt's "Cosmos," and accompanying me to the door, said he was very happy in making my acquaintance, although I had been long well known to him by name. On the following day I called on Lady Franklin with my plan; met Lady Sabine there, who said she regretted not having seen me yesterday when I called.

Wednesday, 11th.—Having received a letter of introduction to Sir Francis Beaufort, from Lady Franklin, I went to the Hydrographical Office and gave it to Captain Becher to deliver to the admiral, who saw me, and on my placing my plan of search in his hands he desired Captain Becher to bring in the Polar chart, on which I traced out my plan, pointing out to him that I believed both Jones' and Smith's Sounds opened into a Polar ocean, as well as Wellington Channel. He appeared pleased with my views, which seemed to coincide with his own, as he said that he considered the broken land at the head of Baffin's Bay formed islands. Sir Francis cautioned me not to be too hasty in laying my plan before the First Lord, until I could get it backed up by some powerful Arctic or other influential authority; otherwise, he said, he was sure it would be laid aside, for the Board of Admiralty only acted under pressure from without. He added, Sir Edward Parry's opinion would have greater weight than any one's, and strongly urged me to see him. Mr. John Barrow was introduced to me by Sir Francis Beaufort, and as I was quitting his office with Captain Becher, Sir Francis advanced towards me and said, "I must have your hand, and hope yet to shake hands with you again, to wish you success on your departure." His reception altogether was most gratifying, warm, and friendly.

On the 12th, at six p.m., I dined with Captain Denham, at No. 21, Charlotte Street, Portland Place. I met him this morning at Lady Franklin's. In my note to Sir Edward Parry, I had asked him to appoint a place and time for my seeing him on his way through town to Norfolk.

On the 14th I saw Captain Denham at his club, the Senior United Service, and showed him my plan of search, after which he showed me round his club, and we afterwards called on Lady Franklin in Spring Gardens, taking Parry's answer to my note with me.

Monday, 16th.—At 9.30 p.m. I waited on Sir Edward Parry at Sir Edward Buxton's, No. 10, Upper Grosvenor Street, Grosvenor Square. Parry received me with a hearty shake of the hand, and said he felt flattered by my referring to him for his opinion, and that mine was a noble offer, and he had no doubt whatever but I would accomplish all that it was possible to do under such circumstances; had the plan come from any one else, he said, he should have at once laid it aside. But he felt satisfied that from my experience, whatever I proposed was deserving of careful consideration; and on giving him both my plans, one being by the Coppermine River to the Polar Sea, he said he would take two or three days to consider them over, and then write me an answer. His own opinion was that Franklin had got beset westward of Melville Island; nevertheless he agreed with me generally in my views.

Wednesday, 18th.—I called on Sylvestre, at No. 96, Great Russell Street, to ask his opinion as to the best-adapted stove for a hut, to winter in the northern regions, with the least possible consumption of fuel. He told me, after giving me all the details, that the framework of a house fitted with a stove could be completed in a fortnight, at an expense of about 150*l.* This evening I dined at the "London Tavern," Bishopsgate Street, to which I had been invited by the directors of the Auckland Islands Company. About 300 sat down to dinner; Admiral Dundas in the chair, with Mr. Charles Enderby on his right. I was seated next to Admiral Sir George Back, a place allotted me by Charles Enderby, by whom I was introduced to the old Arctic explorer, who shook hands with me, and said he had noticed my name on the plate next to his. I shook hands with Sir Edward Sabine, who took his seat just opposite to mine, and next to Sir Roderick Murchison: the latter wore a large glittering star on his breast. I had a long talk with Sir George Back about my plans of search, which he said he entirely approved of; but added that he feared the Admiralty could not be moved to do anything more than already done.

Tuesday, May 1st.—Whilst at breakfast this morning I received a note from Miss Franklin, requesting me to accompany her mother to the Enderbys' at 10.30 a.m. On reaching Lady Franklin's residence, No. 21, Lower Bedford Place, Russell Square, I found a cab already at the door. We saw the Enderbys in the office, whom her ladyship had some talk with, and found that a vessel such as is required for Polar service was somewhat difficult to obtain. But they took her to a shipbroker's where several were offered. Returning through St. Paul's Churchyard, she alighted at Dollond's, the optician, who she said, had been an old friend of her husband's. On her return to the cab she told me that Mr. Dollond had very liberally offered to subscribe 100*l.* towards the outfit of any expedition she might send out. Having transacted all her morning's business, she put me down opposite Craven Street, where I lodged, I having renewed the offer of my services as a volunteer, to defray all my own personal expenses myself, on condition that I should be left to act independently in carrying out my own plans of search.

Calling on the following day at Spring Gardens, Lady Franklin had not yet arrived, but it was satisfactory to hear from Mrs. Leaves that she was glad Lady Franklin had me to accompany her to the city, and "that my noble conduct had won golden opinions for me with Admiral Beaufort." Miss Cracroft adding, "Lady Franklin appeared to think you are the only friend left to the cause she has so much at heart." I have certainly used all my influence with her ladyship throughout to dissuade her from risking her own fortune in such an uncertain and hazardous venture, more especially if without the sanction of her friends; but to leave the whole in the hands of the Government authorities, in the event of her not being able to get up a private expedition by public subscription.

On the 11th I walked across Blackheath to Kent Cottage, the residence of Sir Edward Sabine; found both him and her ladyship at home, and being early in the day, about noon, he was in his dressing-gown. We had a long discussion about the search; I showed them both my plans. He considered that the Wellington Channel was of the utmost importance, and thought the exploration of it should be left entirely to myself. On my return to my lodgings in the evening, I found a letter from the Admiralty on my table, containing an answer in the negative to my proposed plan of search.

Having thus failed in my efforts to prosecute the search for Franklin, I on the 12th of July left town for Dover; and on Saturday, the 14th, embarked in the Belgian railway steamer for Ostend, where I arrived on the following morning; and after breakfasting at the "Hôtel de Swede," took the train to Aix-la-Chapelle, where I purposed remaining a month for the benefit of the mineral waters and baths. The weather was beautiful, a bright sun lighting up the rich and fertile landscape through which the train passed in this highly cultivated country. A level plain of cornfields, meadows, and orchards, divided by hedgerows of the ash, birch, pollard, and elder. Cherries, both black and red, appeared to be the prevailing and most abundant fruits. Windmills stud the surface of the country, giving to the whole scene a very picturesque aspect. Arriving at Aachen in the evening, I took up my quarters at Dremel's "Hôtel du Grand Monarque," in Buchell Street.

My bedroom was No. 68, on the second floor. Breakfast hour, nine a.m. Dinner at the *table-d'hôte*, 1.30 p.m. About thirty-six sat down to table. Tea at six p.m. The number ultimately increased to sixty. I made my customary hour of rising six a.m. The hotel contains 220 rooms, twenty of them sitting-rooms, with forty-two servants, a well-managed and very comfortable establishment.

Tuesday, August 21st.—After breakfast I left for Spa, where I arrived in the afternoon. The approach is very pretty, the road passing between a fine avenue of trees. I took up my quarters at the "Hôtel des Pays Bas," dining at the *table-d'hôte* at four p.m. On the following morning I walked as far as the Souvenier Spring, about two miles distant, along a charming road, to breakfast.

Found large parties at breakfast in the open gravel area between the two springs and the cottage; a lovely spot, surrounded by woods, having paths winding through them. The breakfast was excellent; the best very small white loaves I ever tasted, with delicious fresh butter, capital coffee and rich cream; and all for a franc. A cock and half a dozen hens attended at my table to pick up the crumbs I scattered amongst them. I had a glass from each of the springs, Grossbeck and Souvenier, returning to Spa through the young oak-woods, in which blackberries and other wild berries, both red and black, were growing amongst the grass. During my sojourn here, I visited all the spa waters in the vicinity, and took my morning chalybeate bath. On my departure I purchased some of the beautiful paintings on a light wood, for which Spa is so celebrated.

On the 28th, after having breakfasted as usual at the Souvenier Spring Cottage, I took my departure from Spa by the train. Passed through Liege and Malines to Antwerp, and slept at the "Hôtel de Rubens." After taking a survey of the place, on the following day I embarked in the *Antwerpen* steamer on my return voyage down the river, passing Flushing. On the following morning was off the Nore. Landed at the docks before noon of the 30th, and went to my old lodgings, No. 11, Craven Street, Strand.

On Tuesday, the 11th of September, I started by the Great Eastern Railway on an excursion through the eastern counties for a week; passing through Soham, Newmarket, Bury St. Edmund's, and Ipswich; and down the River Orwell, by the *River Queen* steamer, to Harwich, staying for a day or two at the "Three Cups Inn" there; and my old friend, Commander John Stephen, R.N., residing in the vicinity, I dined one day with him and his family at their cottage, saw his aviary of fine canaries, and accompanied him in some rambles in the neighbourhood.

I had every morning a most refreshing bath in the open sea. On Monday, the 17th, I returned to town by the *Orwell* steamer.

September 20th.—Having made up my mind to furnish a small cottage in the suburbs of London, both for a storehouse for my natural history collection and as a home for one of my sisters, I to-day commenced the search for some suitable place, and ultimately fixed upon No. 11, Apsley Cottages, Twickenham Green.

November 6th.—The *Herald* reported the arrival of Sir James Ross's ships off Scarborough.

Wednesday, 28th.—I called on Captain Denham, who accompanied me to the Admiralty, where we saw Sir Francis Beaufort, from whom I learnt that there would be three chances for me in getting out in the proposed forthcoming search, as the two ships but just returned were to be re-commissioned immediately for Behring's Strait, with the chance of an overland route for myself, *viâ* Panama, to join them at the Sandwich Islands with despatches. The second route mooted, would be by the Coppermine River, on Rae's track; and the third, from the eastward, *viâ* Baffin's Bay, by my own proposed plan. I afterwards called on Lady Franklin, who expressed herself by no means satisfied with the results of Sir James Ross's late expedition, and asked me to draw out a new plan for her with my own suggestions as to whatever now remains to be done in any future search that might hold out any prospect of better success than has hitherto attended the efforts already made. Her ladyship paid me a compliment on my writing, and the soundness of my views, already brought under the notice of the Admiralty in my plans of search.

December 4th.—I called on Mr. John Barrow at the Admiralty, from whom I learnt that there could be no longer any doubt about my being employed in the search. He told me that when the Arctic Committee met he called attention to my name, when it was acknowledged by all present that there could not be a more fitting person for the command of a boat expedition.

Saturday, 15*th*.—I called on Lady Franklin with my new plan of search. Her niece, Miss Cracroft, and herself both read it; expressed themselves pleased with it, and said it was clear, and could not be improved upon; as did also Mr. Barrow, on my taking it to his office. On the 18th I saw Sir Francis Beaufort, when he told me that he took my plan home with him last night, read it, and that it met with his entire approval. As Sir Francis offered to transmit a copy to Sir Edward Parry under Admiralty cover, I left a copy with him. Meeting Mr. Walker, of the Hydrographical Office, outside the admiral's door, I showed a copy to him, when he remarked that it was the most likely project to succeed of any he had hitherto seen.

CHAPTER XVII.

Plan of boat voyage brought before the House of Commons—Plan rejected by the Admiralty—Cross purposes.

Friday, January 4th, 1850.—I called on my old Tasmanian friend, Judge Montagu, with whom I had a long talk about the Falkland Islands, to which the Government had offered him the appointment of stipendiary magistrate, at a salary of 500*l.* a year. But he told me he had not decided on accepting it until he had heard from me my opinion of it, which I fairly gave him, and it was this—that the islands had little to recommend them. I saw Mrs. Montagu and the two children, when he told me that his son was desirous of entering the navy. He expressed so much interest in the search for Franklin as to say that if he had neither wife nor children he would join me in it. He told me that he had once seen the celebrated African traveller, Mungo Park, and had been acquainted with both Denham and Clapperton.

Saturday, 19th.—On going this morning to the paymaster-general's for my half-pay, I learnt from Messrs. Mountenay and Coles, two of the chief clerks in the office, that Sir John Ross wanted me to join him in a search for Franklin. Afterwards, on seeing Mr. Barrow at the Admiralty, I learnt from him that Sir Francis Beaufort had deemed it advisable to delay the laying my plan before the Board until the Behring's Strait ships had sailed; and subsequently reported most favourably on it, with the view of sending me out in Captain Penny's whaler, the *Advice*, he having offered to take out any scientific officer the Admiralty might be disposed to employ in the search; but nothing, it appears, will be done until after the meeting of Parliament.

Wednesday, 23rd.—I received a letter from the Admiralty thanking me for my plan of search, and that if adopted their lordships would communicate with me again. On Saturday, February 2nd, I called on Chisholm Anstey, M.P., at his office in the Temple, when he told me that Sir Robert Inglis was to bring forward a motion on the Franklin expedition in the House of Commons this evening, and if I wished to be present to send up my card to him. Passing by Charing Cross, I met Sir Edward Parry walking arm and arm with Sir John Richardson, to whom he introduced me; we however had met years gone by. We walked together as far as the Admiralty, where I parted from them in the hall. Parry told me Scoresby was in favour of my going out. I called both on Mr. Barrow and on Lady Franklin, to acquaint them with my intention of being at the House to-night. Met Sir John Ross in the Strand, and afterwards Dr. Daniell, who accompanied me to the House of Commons, and obtained for me an admission ticket to the Strangers' Gallery, and at 7.15 p.m. I entered the House for the first time in my life. Mr. Hume, I believe, was speaking on the Church question, followed by Mr. Aglionby, when at 8.45 Sir Robert Inglis rose, and strongly urged upon the House the necessity of a further search, from the eastward, &c., seconded by Mr. Anstey, who brought my own plan of search prominently forward, urging that whatever expeditions might be determined on, the open boat expedition I had proposed up the Wellington Channel and by the sounds at the head of Baffin's Bay—Smith's and Jones' Sounds—should at least not be neglected, and stating at the same time my former Polar services and claims to the command. Sir Francis Baring rose and said that there would be another expedition, eastward, and that the propositions now alluded to would receive every attention. The whole subject occupied about half an hour. I remained about two hours, quitting the gallery at 9.15 p.m., whilst Mr. Horsman was speaking, and I returned home by the 10.30 p.m. train. Blowing a heavy gale of wind. It was two a.m. before I got to bed at Twickenham. My old friend, Dr. Daniell, who met me coming out of the House of Commons, told me he was to sail on Friday next for Africa.

Wednesday, 20th.—To-day I went to the Admiralty with the revised copy of my boat equipment, and after showing it to Mr. Barrow, Admiral Beaufort, and Captain Hamilton, the latter started some difficulties in the way of my going out in Penny's vessel, adding that he wanted me to go out with Captain Austin. I left the plan to be laid before the Board. Afterwards, calling on Lady Franklin, I explained to her that the difficulties raised by the officials at the Admiralty originated in the much-to-be-regretted indecision and entire absence of all concert on the subject at the Board: a most disheartening state of things to have to deal with. Lady Franklin said, "Adhere to Jones' Sound." Captain Peel, R.N., son of Sir Robert Peel, who is also a volunteer for Arctic service, coming in during the discussion, said from an interview he had had with the First Lord yesterday he saw no hope himself of being employed in the search; although as he told me that, like myself, he was quite ready to go out, even if they only gave him charge of an open boat, so zealous was he in the cause. I had a long conversation with him on my own boating affair progressing so slowly and unsatisfactorily, which seemed to interest him much. He is a fine, frank, manly specimen of the naval officer, with very pleasing manners, shaking hands most cordially with me on my rising to leave, as if he had known me for years, instead of a first casual introduction. Poor fellow, like many more I have met in the journey through this uncertain, transitory phase of existence, he is now no more.

March 1st.—My offer to be employed in the search was finally rejected by the Board of Admiralty, after all my unwearied application, and the unceasing and able support it had met with from the hydrographer, Sir Francis Beaufort, who said he was as much astounded as myself at the result; more especially, as he told me Captain Hamilton, the secretary, had complimented him for backing up my plan by a report on it himself, and even going so far as to get a model of a north-country boat for me.

On making my failure known to Lady Franklin she expressed great surprise, and warmly shaking hands with me, said, "I never forget an old friend," stating that she would see that I went out some way or other, if at her own expense. Admiral Beaufort, Captain Hamilton, and Mr. Barrow all urged me to see Mr. Anstey, M.P., and get him to bring my case under the notice of the First Lord, as they seemed to be of opinion that the refusal was an act of the Board in the absence of the First Lord, and without his knowledge. I consequently saw Mr. Anstey, at his chambers in the Temple, and made this request of him, pointing out at the same time the injury that would accrue to myself if not employed, after things had gone so far, and that the only reparation and compensation that could be made me would be my promotion to the deputy-inspectorship for past services in the Antarctic expedition, which had been so long in abeyance. Anstey replied he would see the First Lord himself. He not only saw Sir Francis Baring, but also wrote some very strongly-worded letters to him, of which he kindly furnished me with copies.

After repeatedly urging my employment, both on the First Lord, Admiral Dundas, and Captain Hamilton, and encountering every kind of evasion, he at last wrung out a reluctant and indefinite promise that I should be sent out; which, however, after all was not fulfilled by them. Notwithstanding, I used all the pressure I could bring to bear on the First Lord. I saw him again on his *levée* day, entered into a full explanation of my plans, reminding him that Admiral Dundas had told a member of Parliament, Mr. Anstey, that I was to have charge of a boat expedition, and that Captain Hamilton, the secretary to the Board, had given me every encouragement to hope that I should be employed. To which Sir Francis replied, "I have read over your plans, which appear to me to be attended with great risk;" and on my attempting to remove this impression, he said, what he more particularly alluded to was the want of some place of refuge for me and my boat's crew after being left by the ship to our own resources, which I met by stating that there were abundant depôts, all accessible, to fall back upon, &c. He shook me by the hand on rising to go; but all I could elicit from him was, "You will hear from the Admiralty in a few days." This was on the 12th, and yet, on going into Captain Becher's office on the 28th, and meeting Captain Austin there, he, after shaking hands with me, said, "I thought you had cut me, not having seen anything of you of late; but although, having been so busy myself, I have seen nothing of your plans, yet I should have had no objection to your going out with me, and hope that you have not thought so."

Lady Franklin also told Mr. Anstey that Penny had assured her he had no objection to a naval officer going out with him, provided that he went out independently, neither placed over him nor serving under him. Be this as it may, my kind friend, Mr. Barrow, acknowledged to me that I had been very ill-treated throughout, and that from first to last there had been no disposition on the part of the Board to forward my plans. Indeed, all that had been done was through the hydrographer; and the whole truth was that Admiral Dundas had from the first believed that Franklin's ship had never got out of Baffin's Bay, but had been lost there, and he consequently cared nothing about the prosecution of the search, and the Board would give themselves no further trouble about it than public opinion and the pressure from without compelled them. I subsequently learnt from my former boating companion, Commander Phillips, that Penny had actually refused to give me a passage out in his vessel, notwithstanding both Sir Francis Beaufort and Admiral Dundas were reported as having said to him, "You must find room for McCormick, and take him out with you."

The known antipathy existing between the hydrographer and the Board was unfortunate for myself, after all my own exertions in this humane cause, and which met with nothing but the coldest indifference, nay injustice, from the Board of Admiralty. One satisfaction is, they cannot deprive me of the good opinion of my friends and supporters—that I could not have done more than I did.

On the 10th of April, after hearing from Mr. Barrow that Captain Hamilton had now relinquished all hope of my being sent out, I called at Admiral Beaufort's office, to thank him for all the kind support he had given me. He received me very kindly, and said, "The Admiralty have acted very unjustly towards you, but your reputation will stand as high as ever throughout the length and breadth of the land;" adding, "You have from first to last perseveringly urged your plans in the most straightforward, generous, and handsome manner, which every one must feel."

On my telling him that Lady Franklin had offered me

the command of a private expedition of her own, to search Ross's Strait, and much wished me to undertake it, he replied, "Do not pledge yourself to anything of the sort. Franklin will not be found there." Remarking, "I have kept out of her ladyship's way of late, in consequence of this feeling, although I feel that I have been generous in doing so, as I am anxious to have that portion surveyed, under the impression that there is a passage into the bottom of the Gulf of Boothia." The admiral throughout has been my most staunch friend.

On seeing Mr. Anstey, on the 13th of April, he told me that he had had several conversations with the First Lord and Admiral Dundas since he saw me last, and that both had acknowledged to him that mine was a hard case, and that the whole had been, by his own admission, bungled from the first. He added that I could go out in any one of the ships I liked, but that the difficulty was about a boat's crew, for which there was no time left now. Yet Anstey wanted to persuade me, and in the most emphatic manner, that it was impossible for me to stand higher than I did in the opinion both of Sir Francis Baring and of Admiral Dundas. The latter, he said, spoke of me and my plan in such high terms as to acknowledge I was deserving of a ship to myself for carrying it out, had there been time left, and that the main difficulty appeared to be the finding a vessel for a *point d'appui*, for me to fall back upon. Anstey said he told the Admiral that my going out as he had suggested would not be carrying out my own plans, but his, which the Admiral at once admitted. Finally Anstey assured me, that from what passed if I was not sent out they would give me my promotion.

Wednesday, December 4th.—Having written an article on "Christmas Day at the South Pole," which my friend Hunt, the editor of the *Daily News*, had asked me for, as a contribution to the Christmas number of *Household*

Words, for Charles Dickens, I took it to him, when it was at once sent to press, and appeared in that number.

The year 1851 passed away with but few incidents of interest. The searching expeditions returned from the Arctic Seas about Michaelmas, without having accomplished anything of importance, but quarrelling amongst themselves; Austin and Penny at loggerheads, their jealousy of each other causing obstructions to the service, and thus defeating the very object for which they were sent out. There was also a schism between old Sir John Ross and his able and experienced ice-master, Abernethy; and altogether it was a great mistake to pack so many vessels together in one locality, as in the Wellington Channel, where such discordant elements for dissension existed, instead of extending the exploration over a wider area, which would have afforded each ship a separate line of search, and by thus going over a greater extent of ground, so would the chances of finding some clue to the fate of the missing expedition have been increased.

The feeling in the squadron was that Austin was considered to be a goodnatured man, and spent much of his time in giving entertainments, but was also active and bustling in the superintendence of the departure of the sledge-parties.

I was told that the feeling in the squadron with reference to my own plan of search was that it was a very perilous one; the same thing was told me by Sir Francis Baring. Notwithstanding which, as rumour stated another expedition would be sent out in the ensuing spring, I lost no time in again offering my services to the Admiralty in a letter to the Board dated the 28th of November, to which I received a reply in acknowledgment on the 2nd of December. And on my mentioning to some of my friends that if I failed this time in moving their lordships to employ me, I would endeavour to get up a private expedition by subscription, I soon had offers of subscribers to begin the list with from my friends John Barrow, Dr. Richard King, Messrs. Gillott, the uniform makers to the Admiralty, and others, and very probably might have been enabled to carry it out, had not the Admiralty at last given me a chance, such as it was.

CHAPTER XVIII.

My new plan of search—Difficulties with officialism—Am at last offered the appointment of surgeon to the *North Star*—Accept it on urgent pressure of friends—More obstacles—Departure of expedition.

Wednesday, January 21st, 1852.—Having written out a new plan of search, called for by the changed aspect of things, I took it to Sir Francis Beaufort, who after reading it over carefully seemed much pleased with it, and turning to me said, "This shall appear in print some fine day; it will show that there are others who think deeply on this subject, and the Admiralty must think your remarks years ago on Cape Riley and Beechey Island a striking coincidence;" adding, "How did you then come to name these places in particular?" He was much interested with the part relating to the means of sustenance for the missing crews pointed out by me, and said, "The public should know this." He, however, advised me to reverse my plan of search, and propose going out to the Wellington Channel direct, to commence exploration in a north-east direction, instead of to the north-west from Smith's Sound; otherwise, he said, it would be sure to be thrown overboard by the Admiralty, on the ground that Smith's Sound had never been seen clear of ice. If I made this alteration, and enclosed it to him, he said he would himself put it into the hands of Captain Hamilton, and give it all his support, although he could not promise that it would be successful. On Mr. Barrow's coming into the office he repeated to him most of what he had said to me. After thanking him for his kind support, I took my plan with me, to draw it out afresh. On the following day, the 22nd, I sent off my corrected plan by post.

Saturday, 24th.—On seeing Mr. Barrow at the Admiralty, I learnt from him that he had seen nothing of my plan, and that unless Admiral Beaufort got it early before the Board it would be too late for the printed returns to the House of Commons. As this was a most essential point for its success, I saw the admiral at once about it, when he said, "I put your plan into Captain Hamilton's hands myself;" and on my stating Mr. Barrow's apprehensions about it, he replied, "I'll see to that, and get it through." On leaving the Admiralty I met Sir George Back, who took my arm, and walked as far as Waterloo Place with me. Entering into a friendly chat on the Arctic search, he seemed to cling to the notion that the coast about Pond's Bay had not been sufficiently searched. On his asking me what my present plan was, I told him I would send him a copy to his residence, No. 106, Gloucester Place, Portman Square. On leaving him I fell in with a former messmate of mine some twenty years ago, now Captain Chambers, R.N., who told me that my plans had been much talked about at the clubs, and that the feeling was I had been most unjustly and shamefully treated. On the 28th I was introduced by Mr. Barrow, in his office, to Commander Maguire, going out to command the *Plover* at Behring's Strait—a frank, jolly-looking sailor, who shook me warmly by the hand, and said, "I am most glad to make your acquaintance." And on my plan being named he said, "Could you point out the *very spot* where you could put your hand upon Franklin they would not listen to it at the Admiralty, everything being left to the Arctic Council."

Of the truth of this indeed I have had dire experience. On Mr. Barrow telling me that he had seen nothing of my plan, and Captain Hamilton had gone to Brighton, I said I would see the hydrographer about it; but he told me Sir Francis Beaufort would not go near any of their lordships, as they did not pull together. My letter of the 27th ult., it appeared also, he had not been able to get, and on referring to the large book of names, found that my own name was not therein entered. I told him that it should not be *ignored* in that way, for I would bring up a copy I had by me, with their lordships' answer acknowledging the reception of it, to-morrow, for insertion in the Arctic returns. And on the following morning I took it to his office, and wrote a note at his desk, authorizing him to make use of this copy in printing the returns.

February 2nd.—I again called on Mr. Barrow, and learnt from him that he had succeeded at last in getting my plan by going himself to Captain Hamilton, and saying to him, "If you do not let me have it, I know it will break his heart if it is not printed with the returns to the House of Commons." This was a most kind action on the part of my friend Barrow, and will not easily be forgotten by me.

On the 5th I had a long conversation with Captain Hamilton, having gone into his office to see the models of an east coast of England coble and Parry's sledge. He made some comments on my plan, which he expressed his approval of, but thought I had dwelt rather strongly on the swell which sets out of Smith's Sound. But this was one of the strongest arguments—together with the very large floes of ice which were annually discharged from it, beyond all conception as coming from a mere sound—on which I had formed so decided an opinion, and come to the conclusion that whenever explored it would be found to open into a vast Polar ocean, forming one of the outlets for the passage of the enormous floes on the breaking up of that ice-encumbered ocean. Captain Hamilton expressed a strong desire that I should set the question at rest with reference to Jones' Sound, which it appeared to him had been only very superficially examined, and thought it might be done from the Wellington Channel by an overland journey, as he was not satisfied about the cairn said to have been seen by a whaler in Jones' Sound.

On the 10th I met Sir Edward Belcher (who is to command the forthcoming expedition) in Barrow's office, and learnt from him that he had seen Admiral Berkeley about me, and had proposed my going out in his ship as surgeon, with another surgeon under me for the ship's duties. To this, as I was circumstanced, there were insurmountable objections in the position in which I stood with the director-general of our department, who it was well known was no friend of mine; and therefore to have any chance of carrying out my plans efficiently it was of the first importance that I should go out entirely unfettered by the medical department, by being appointed as additional surgeon for boat service, which would relieve me of all professional responsibility. To this he replied, "You had better see Sir Francis Beaufort on the subject;" which I at once did, and found him engaged with Captain Washington, the future hydrographer, to whom he introduced me, and said that I had been many years a volunteer for this service, and that the whole of the coast round Cape Sir John Franklin might be searched by me; and both agreed with me that I ought to go out "additional" for boat service. I lost no time in making this known to Sir Edward, who said he was quite willing to have it arranged as the admiral wished, but thought that objections would be made officially to such a change in the general routine of the service.

The 17th was the First Lord's *levée* day, for which

I had entered my name and was about attending, when Sir Francis Beaufort advised me to pass it over, as in the present state of things it could do no good, and might do harm. I therefore had the pen run through my name on the list.

On the 19th I accidentally met my old commander, Sir Edward Parry, coming out of the Admiralty. He shook hands, and on my telling him that I was to go out at last, he said, "You will have plenty to do, and you will do it." I met Sir Edward Belcher in the hall of the Admiralty, when he told me that Captain Milne would not appoint me till the ship was ready for sea. These few words carried much meaning with them, and prepared me for the persevering obstruction I was destined to encounter in the carrying out of my long-cherished plan. On the following day I again met him, when he told me in a hurried, excited manner that I could only go out as surgeon of the *North Star*, and that an assistant-surgeon would be appointed to relieve me on going away in my boat expedition. This most unexpected announcement was excessively mortifying to me, and on seeing Mr. Barrow and Sir Francis Beaufort about it they both strangely enough, acquiesced in it, notwithstanding all that had passed only the other day. But I suppose they saw there was no other chance for me. On the 21st I met Sir Edward Belcher in Barrow's office, when he asked me if I had come to any decision, as either a surgeon or an assistant must within ten minutes be appointed to the *North Star*. As I still held out for an additional appointment unfettered, he ended by saying that he had brought it before their lordships a dozen times, and that making difficulties would only injure the whole cause; and it was ultimately agreed to refer the whole to the hydrographer. On my entering his office, Sir Francis Beaufort said, "Sir Edward has offered to give you an assistant-surgeon, so there can be no obstacle in the way of your leaving the ship in your boat." Sir Edward himself then asked me whether he should ask for my appointment to the *North Star* or not. On my reluctantly giving my consent, Sir Francis Beaufort, after Sir Edward left the office. said, "I was afraid, after your peremptory manner yesterday, that you had declined going out under the existing circumstances, and I thought you were mad in so doing." Adding, "I have had a high opinion of you. I like a man with a heart. You have proved that you have got one; and *I* want you to go out." Thanking the good old admiral for his good opinion of me and friendly support, I left the office.

Tuesday, February 24th.—I went to Chatham, and on board the *North Star*, selected my cabin, the after one on the port side; took up my commission to her at the dockyard, and wrote a letter to the Admiralty, reporting my having joined, and another recommending Sylvestre's stove for warming the ship.

On the 4th of March, whilst I was at the Admiralty, a young medical student came to offer himself as a volunteer for the expedition, and on my inquiring into his qualifications, it came out that he was a son of an old fellow-student of mine at Guy's and St. Thomas's Hospitals, by the name of Woodman, and although I made every effort to get him appointed, failed in doing so; and there can be no doubt whatever that the obstacles thrown in the way of an assistant-surgeon's appointment arose from the desire to deprive me of one, so that when we got out the plea of medical duties might be brought forward to keep me on board the ship; and ultimately I only succeeded in getting one, lent from Sir Edward Belcher's ship at the last moment, by my telling Admiral Beaufort that I was firmly resolved, even now, not to go out unless this point was satisfactorily settled before the ship sailed. I strongly urged on Sir Edward Belcher the appointment of Mr. Woodman to the *North Star*, stating his high qualifications, and the acquisition he would be to the service. His reply was that he would have done so himself to oblige Mr. Bransby Cooper, who had recommended him to him, but for the director-general, whom he had seen about it, and who would not hear of a new entry into the service—and he might have added, so far oblige me. On my seeing Mr. Woodman, who had been very anxious to get out with me, and acquainting him with the result of my efforts to serve him, he told me that on naming me to Bransby Cooper, the nephew of Sir Astley, he remarked that I was going to destruction in attempting to carry out such a harassing undertaking. This feeling is very much in keeping with what Captain Austin professed on my first meeting with him, after his return, in the Arctic Council room. On offering his hand he confessed that he had objected to my going out in his expedition on the ground of his unwillingness to incur the responsibilty of dropping my boat and party to take care of ourselves when wholly cut off from the ship; humanity, he pleaded, justified him in so doing. This was all very plausible, but entirely wasted upon me, as I replied, all this had been fully considered by me long ago, and that I thought he must himself now feel convinced that all his groundless fears for myself and party would not deter me from carrying out my plan, even without his aid; adding, that whatever I undertook after mature resolve nothing could prevent me from carrying out, come the time when it would, albeit beset, as at the present moment, with like opposition and obstruction.

On the 15th Mr. Barrow, whilst I was in his office, kindly sent his messenger to get my name entered in the First Lord's *levée* book before it came down into the hall, and, even then, was just in time to be the last name in the list for the first *levée* day, next Tuesday week. He asked me to make a sketch of Disco for him, which I promised to do, and to send him home.

Tuesday, March 23rd.—At four p.m. I attended the First Lord's *levée.* On entering the room the Duke of Northumberland rose from his seat and requested me to be seated in a chair opposite to himself. I put into his hands a copy of my letter to the Admiralty dated the 6th of June, 1850, containing a record of my services, with some testimonials, stating, at the same time, that being about the last on his grace's list, I felt that I ought not to take up his time at such a moment with a *vivâ-voce* statement of my claims, if he would permit me to leave this record with him, which he kindly received, asking me what it ended with, and I replied, my promotion to the deputy-inspectorship; adding that my naming my promotion at the present moment was solely that his grace might bear in mind, on the return of the expedition, when the usual general promotion of the officers took place, that I rested my claims to promotion on special services which I had long ago hoped would have obtained for me special promotion to the half-pay list. He replied, "I will have it all recorded." I also reminded him that I was simply appointed surgeon of the store-ship *North Star*, an appointment beneath my position and standing in the service, and in every way calculated to fetter me in following up the great object of my heart, the discovery of traces of the fate of my old ships, the *Erebus* and *Terror*, and their ill-fated commander, Sir John Franklin, who had been an old and esteemed friend of mine; and I had accepted the position with all its drawbacks rather than be deprived of having some share in the search. It was also the express desire of the hydrographer, Sir Francis Beaufort, who had so warmly supported my plan, and done all in his power to obtain for me an independent and special appointment for carrying it out. The whole interview did not occupy more than five or six minutes. I rose to go, when the duke rose from his seat, and accompanying me to the door of the room, said, "I wish you success; you go with the best wishes of the country with you;" shaking me warmly by the hand as he said so.

On entering Mr. Barrow's room afterwards, I learnt from him that Sir Edward Belcher had been making a lame and futile attempt to get himself and his friend, the director-general, out of the dilemma, in reference to the appointment of an assistant-surgeon to the *North Star*, by telling Mr. Barrow that he had been under the impression that the director-general intended appointing one, and afterwards found that he had left it to him to do so. The fact was, both had been juggling together to defeat me but failed, so far as an assistant was concerned.

Wednesday, 24th.—I met Sir Edward Parry at the Admiralty, who told me I ought to be taken up the Wellington Channel in one of the ships of the squadron, and landed at Cape Sir John Franklin, or I should lose much time in getting over ground already gone over in Wellington Strait; and promised me that he would communicate with the First Lord on the subject. He also told me that it was his wish that a vessel should be sent in the direction of Melville Island in search of the *Investigator*.

On the eve of my departure on a somewhat hazardous enterprise, I had deemed it only prudent to see my mother and sisters all settled under the same roof; and having lately written to them to prepare for a removal to my cottage at Twickenham, on the 25th I went to Dover by express train, but had some difficulty in rousing them to muster energy enough to set about an exchange of residence. However, I got some of the chattels packed, and returned to town on the following day. On April the 6th I went to Dover again, and the next day, having seen the remainder of the luggage to the goods office, I brought my mother and sisters, all invalids, away with me in a first-class carriage to ourselves by the train. Took a coach from the London Bridge station to St. Paul's Churchyard, and from thence by the Chertsey omnibus to Twickenham Green, which we reached at 6.30 p.m.

Saturday, April 10th.—I met Dr. Rae in Barrow's office, to whom I was introduced. We had a long chat together over the Arctic chart, on which he pointed out to me a spot to the southward and westward of Cape Walker as, in his opinion, the most likely spot for discovering the fate of Franklin. He said he had read all my plans, and had expected that I should have been sent out long ago. On the 13th, my friend Barrow having promised to accompany me on board the *North Star* before her departure, we started by train for Woolwich. On our arrival at the station we found a group of officers, with Captain Kellett, about going to the hospital of the Royal Marine division, on a survey of an old ice quartermaster who had been very injudiciously, not to say officiously, during my temporary absence, reported unfit for the service. So that my timely arrival very opportunely saved the poor old fellow from what would have been simply an act of injustice to him; for examining into the causes of his rejection I found nothing whatever in justification of the step taken, and reported him fit for the service, the medical officers of the hospital entirely agreeing with me. The survey was accordingly annulled, and I took him in my own ship, and he turned out one of our most reliable petty officers.

Saturday, 17th.—I took my leave of my excellent friend Sir Francis Beaufort, Mr. Barrow accompanying me into his office. The admiral produced the Polar chart, and called my attention to the cairn which had been seen in Jones' Sound by a whaler, and expressed a wish that I should search for it. He gave me two copies of his "Sun-tables," shook me heartily by the hand, and wished me success. Quitting his office I met Captain Hamilton, and he made the same suggestion to me the admiral had done, adding, "I have great faith in you, and hope you will not forget to search for the cairn in Jones' Sound, or not see my face again."

The same afternoon I called at No. 21, Bedford Place, Russell Square, to take leave of Lady Franklin and her niece, Miss Cracroft. Her ladyship also dwelt strongly on the necessity of a search in Jones' Sound for the cairn. I remained talking with them for above an hour on the unsatisfactory position in which I was placed, and the keen disappointment it was to me, after all my exertions in the cause, to have been denied an independent command, but at the last I had determined to set all obstacles aside, and go out and do my best. On bidding her ladyship good-bye, and shaking hands with her, she pressed me to stay to dinner, and on my declining, asked me for to-morrow, which, for the same reason, want of time, I was also compelled to decline.

Monday, 19th.—This being the day for their lordships' inspection of the ships, accompanied by a friend I entered the train at the London Bridge terminus. Having met Captain Austin on the platform, he joined us in a first-class carriage to Greenhithe, entering into a long conversation about the expedition, and saying it would be a dashing thing if we completed our work and got home again the same season. He appeared in excellent humour, offering us each a part of the *Times* he was reading. It was 1.20 p.m. when I got on board, and found that their lordships had only left about ten minutes, which I confess was a matter of no great regret to me, after the treatment I had so recently received at their hands. Frank Y. Toms, the assistant-surgeon, whom I had at last succeeded in getting appointed to the *North Star* at the very eleventh hour, had lost no time in establishing himself on board, and I found him in the gun-room—a promising-looking young officer. After showing my friend round the main-deck, the gun-room, and my cabin, we returned on shore, and McClintock came on board just as I was leaving. On our return to town, learning that the officers of the expedition were expected at Beard's photographic establishment to have their likenesses taken, we called in and found Alston, the mate of the *North Star*, getting his taken; but the light proving bad, it was a failure, and both he and I were asked to come again to-morrow.

On Tuesday, the 20th, went to Beard's, in King William Street, and at noon had my portrait taken hurriedly, in a dress uniform coat, in company with a friend, who also sat for one. The light was good. Called at the Admiralty afterwards and bid my friends Mr. Barrow and Captain Hamilton good-bye. Returning to Twickenham, I took a hasty leave of my mother and sisters there, and on my return to town again I started by the last train from London Bridge at ten p.m. Sir Edward remarked as he passed along the platform, seeing me looking out of the door of the carriage in which I was seated: "You will lose your passage." My rejoinder was: "I only share the risk of doing so with the commander of the expedition." The commodore was going by the same train. I got on board my ship about midnight.

As the narrative will appear in another part of the work, I shall only state here that I returned in the *Phœnix* steamer to England, arriving in the Thames on Thursday, October 13th, 1853.

CHAPTER XIX.

Again on half-pay after my return in 1853—Complete my chart at the Hydrographical Department of the Admiralty and circulate a few private copies of my Boat Voyage amongst friends to the search —Promotion still in abeyance—Dr. King's Arctic lectures.

On my return from my Arctic voyage, I took up my quarters at my old lodgings, No. 6, Woodland Terrace, on October 15th, and gave my Esquimaux dog, Erebus, in charge of my former shipmate, Abernethy; my other dog, Terror, having died on board the *Phœnix* on the passage home. On the 17th, I went to the Admiralty and saw my friend Barrow, who welcomed my return with a hearty shake of the hand. He told me he had read my narrative, and said it was the most dangerous expedition of any of them. He asked me to write my autograph on the fly-leaf of a copy of his father's book "Arctic Voyages of 1818," containing autographs of all the officers who have commanded expeditions, and to be deposited after his death in the British Museum. Mine was the last on the page. As I was coming out of the Admiralty I met Sir George Back, in company with another old admiral and another gentleman. He was most hearty in his congratulations, taking my arm and walking with me as far as the Senior United Service Club, all of us chatting about the expedition. Afterwards I called on my friend Dr. King.

Sunday, 23rd.—I spent the day with my old friend Professor Owen, at his residence, Sheen Gate, Richmond Park. He showed me round his grounds, and his little farmyard, with his two cows. Mrs. Owen had gone to make a call on her neighbour, the Hon. Mrs. Liddell, and on her return said Mrs. Liddell wished to be introduced to me; so I went over with Professor Owen, and saw the ancient ash of Queen Elizabeth's time, its hollow shell of a trunk still possessing sufficient vitality to send forth some green branches. Mr. Broderip, the amateur naturalist and distinguished barrister, had been asked to meet me at dinner, and gave me a most cordial reception, shaking me by the hand over and over again. There were two other gentlemen at table, and Mrs. Owen. Mr. Broderip and I returned in a cab to Piccadilly, where I took leave of him, and it being too late for the train to Woolwich I had to walk over Blackheath home.

November 4th.—My friend Barrow asked me to give Stephen Pearce, the Arctic portrait painter, a sitting for my portrait, for his collection of the portraits of commanders of Arctic searching parties. On the 5th, when at the Admiralty, Captain Becher returned me my chart, on which the hydrographer, Sir Francis Beaufort, had done me the honour to substitute my own name for the bay I had in my chart named the Bay of Refuge. On the 7th, Captain Becher introduced me to Mr. Arrowsmith, the distinguished publisher of charts, who gave me an invitation to breakfast with him.

I was employed on the 16th and 17th correcting the compass bearings and variations of my chart of Wellington Channel, and took it to the Admiralty on the following morning and remained all day in the Hydrographical Office, making two tracings of it in Walker's office; and was but just in time to get the alterations made in the new Admiralty chart, as 150 impressions were commenced, of which eighteen had already been struck off from the plate, when Mr. Walker stopped them till

I had seen Admiral Beaufort, and pointed out to him that Mount Franklin had been placed out of its proper position, as there was no mountain whatever in that direction. Arrowsmith himself, fortunately, came into the office as Captain Becher and I were discussing this point. He quite agreed with me that there was no peak whatever in the position laid down in the Admiralty chart, and said that he had left it out in his own chart. He moreover added that it was very dangerous to ships placing such a peak where none really existed. He thought my own peak would be found to agree with the one termed Barrow's Monument. Arrowsmith said that my outline of the coast appeared to him so much more satisfactory, that if I would furnish him with a tracing of it he would introduce it into his own new chart. Mr. Walker then gave me a copy of the new chart, to reduce the scale of my own to it.

On the 21st I again went to the Hydrographical Office, and there traced out my coast-line on a reduced scale for Arrowsmith, and a copy was also given to the engravers, who were set to work on the Admiralty chart to insert it. Whilst I was in Captain Becher's office Arrowsmith came in, and I gave him the tracing from my chart. We had a long conversation on the position of Mount Franklin; he expressed his entire satisfaction with my coast-line, telling me that he would send me a copy of his chart in three or four days, and I promised him a copy of my own narrative in return. During the conversation, Mr. Walker brought in the first impression from the new Admiralty chart plate with the insertion of my line of coast complete. Barrow brought in the sketch-book, with views of the headlands in the Wellington Channel, &c., which I had given him, and he paid me the compliment of having had handsomely bound up, in roan-coloured Russia, with his own crest (the squirrel) and my name in gilt on the cover. My chart and discoveries, with another of Baffin's Bay, were bound up with it.

December 2nd.—I paid my subscription of 2*l.* for my lamented friend Bellot's testimonial, to the Geographical Society.

Monday, 12th.—I called on Arrowsmith, who presented me with a copy of his new chart, the first impression having my own name inserted in the upper corner, amongst the Arctic discoverers. On the 15th, Mr. Barrow gave me a note of introduction to Mr. Scott, the manager at the Queen's printing-office, to enable me to see my boat narrative and sketches through the printers' and engraver's hands. On Mr. Scott giving me Mr. Martin's, the engraver's, address, I at once called on him. He said my sketches were very good, the best he had seen from any of the expeditions, and were deserving of being transferred to better paper than the Government foolscap, and that the *Forlorn Hope* running under the lee of the icebergs would make an excellent picture. Mr. Waldick, one of the artists, being himself a scene-painter, said he would like to get up a panorama from my rough sketches.

February 18th.—I called at the Queen's printers, and at Day's, the engraver's, for my chart, which being finished I received ten copies. I presented Arrowsmith with one, and had an hour's chat with him; he seemed much pleased with it, and said it could not have been better done, and that he believed my coast-line would come out from my dead-reckoning as near as possible to correctness.

On Tuesday, the 21st, when dining at No. 7, New Street, Spring Gardens, with my friend John Barrow, I saw his fine large painting of the Arctic Council, painted by Stephen Pearce: all excellent likenesses, Sir Francis Beaufort especially—it was to the very life, both in expression and attitude. Besides his mother, Lady Barrow (Dowager), and his sister, he had his brother, Sir George, and Lady Barrow, with Mrs. Batty, his other sister, and her two daughters to meet me. The con versation, animated, on Polar adventure, north and south.

On the 25th of March I succeeded in getting the late coxswain of my boat, Rands, into the "*Talbot*," going out this spring to the north; her commander, to whom I had strongly recommended him as ice quartermaster, promising me that he would take him. Captain Washington, the hydrographer (elect) to the Admiralty, travelled in the same carriage of the Woolwich train up to town with me. He thanked me for the copy of my narrative I sent him the other day. Said that the illustrations were a great addition, that one was worth whole pages of written description, and he wished that I could inocculate the service with a taste for sketching; adding that he was much interested in the perusal of my narrative.

On the 28th, I called on my former advocate in the search, Chisholm Anstey, at his office, No. 3, Hare Court, Temple. Found him engaged with a Mr. Crauford, M.P., to whom he introduced me. On my presenting him with a copy of my narrative, he looked over the illustrations, with which he seemed much pleased; Cape Spencer, in particular, he said, was beautiful. He related to me how the commanders of the late expeditions had been cut up in the new quarterly. When I told him, that had we carried out our project at the time he was lending me his aid—some three or four years ago—the Wellington Channel and Polar Sea would now have been our own, his reply was, he had found Admiral Dundas too hard to deal with. Afterwards I attended Professor Owen's lecture at the College of Surgeons; in which the Professor, on commencing the history of the class mammalia, paid me the compliment of introducing my name in a flattering allusion to the late adventurous voyage I had made to a high latitude, and met with traces of the musk-ox, &c.

April 6th.—I met Charles Enderby most unexpectedly in the Poultry. He returned to England, he said, about a week since, and that the whole Auckland Islands scheme had been a failure through dissensions in the company. But he told me he liked the islands, that they were both temperate and productive, the soil rich, with fine harbours; in short, fully answered to the account I had given of them in every way. This was satisfactory.

On the 27th I received a very gratifying letter from Mr. Grinnel, of New York, who has been such an enthusiastic supporter of the Franklin search. On showing it to my friend, John Barrow, in company with whom I dined to-day at Kensington Palace Gardens, with his brother, Sir George, and Lady Barrow and their family, he suggested that I should publish it, and it will be found with one from Judge Kane, Dr. Kane's father, and two letters from Chisholm Anstey, M.P., to the First Lord of the Admiralty, Sir Francis Baring, in the Appendix. The celebrated architect, Decimus Burton, who was one of the eleven who formed the circle at the round table at which we dined, gave John Barrow and myself an invitation to dine with him next week. On learning from Barrow that a promotion was coming out, I urged him in the strongest terms to apply for that of my old assistant, Toms, which he did.

May 3rd.—I received a very flattering letter from Judge Kane, the father of the Arctic explorer. I called on my friend Hunt, the editor of the *Daily News*, at his club, the Reform, and he told me that he had seen Bernal Osborne, the secretary of the Admiralty, who had promised him that he would recommend me for promotion, after many attempts at evasion such as, "There were others with longer services," and that "*No* promotion was given for scientific services;" but he acknowledged that my character stood very high at the Admiralty. Hunt said, "I wanted him to get your promotion for your services, and not as a favour to myself, and I think you will yet get it." However, whatever Bernal Osborne may say, I stand the senior on my list, and have also long medical services to back up my scientific ones. Hunt seemed sanguine about my promotion, and that it rested with Osborne and Sir James Graham, both of whom were to be influenced by political motives only. I replied, that having being so often disappointed I was less sanguine; and in the sequel I was right, for on my return to my lodgings in the evening I found an official letter from the Admiralty, containing a negative reply to mine, and accompanied by a note from Captain Hamilton, expressing his regret at my continued disappointments.

June 2nd.—My attention was called by a brother officer to a very favourable review of my "Boat Voyage," in *Fraser's Magazine* for the present month.

Friday, 23rd.—I removed my mother and sisters from Twickenham to Inkermann Cottages, Norbiton. Had a very busy day in getting the furniture and things there by van, and which took up my time for the next day or two in getting in order.

July 28th.—I took a copy of my "Boat Voyage" to Parker's, West Strand, for the editor, in acknowledgment of the review of it in *Fraser*. On the 31st I called at the Queen's printer's, and settled with Mr. Scott for the extra copies—seventy-three—I had struck off on quarto wove paper, twenty-five copies of which I had bound up in blue cloth, for presentation to friends who had patronized my voyage; this being all at my own expense.

Tuesday, August 1st.—At ten a.m. I went to Stephen Pearce's and had my last sitting for the portrait for Barrow, which occupied three hours, and I left at 1.20 p.m.

On the 5th I received a copy of Creswell's "Views in the Discovery of the North-West Passage," from Day's, the publishers—a very handsome presentation.

On Sunday, the 6th, I accompanied my friend and brother officer, Dr. M'Bain, to the Zoological Gardens, to see the new aquarium. There we met our mutual friend Dr. Bowerbank, with whom we dined in the evening at the Grove, Islington, and saw his museum of fossils, &c. Sir John Herschel was pointed out to me in the Zoological Gardens, and this was the only time I saw our celebrated astronomer. On the 10th I forwarded bound-up copies of my narrative to Lady Franklin, Mr. Grinnel, and Judge Kane.

Saturday, 12th.—I took the train from Euston Square Station to Oundle, on a visit to my old friend, Dr. Clark, of Weldon, one of the few survivors of the battle of Trafalgar. At Oundle I found a chaise ready to take me on to his residence at Weldon. On the 17th I dined at the rectory with the Honourable and Rev. Mr. Hatton, and Lady Louisa Hatton—eleven sat down to table, including Dr. Clark and myself. Having passed three weeks very agreeably with my old friend in a round of drives in the vicinity of the village, I left for town on the 2nd of September, and went to Surbiton to see my friends.

October 10th.—The ships having returned from the north, and a promotion in contemplation, I called on my friend Barrow, at the Admiralty, and had some talk with him as to my own chance of being included in it. His opinion was that he feared it would be referred, as a matter of course, to the director-general, and but for him he himself felt convinced that the Admiralty would have promoted me long ago; and his advice was I should write a letter to the Board of Admiralty, and, if possible, put it into the hands of Admiral Dundas, the medical lord, which I did the next day; and on the 12th, on seeing Captain Hamilton, learnt from him that the Board would not promote me for Arctic service, considering that the only claim to the deputy-inspectorship was for medical service and recommendation by the director-general, who being my known and acknowledged enemy, seemed very like injustice.

On the 13th I received a very handsome letter from my dear old friend Sir Francis Beaufort, and on showing it to Mr. Barrow, and telling him the result of the interview I had with Captain Hamilton yesterday, he said the director-general's treatment of me was most shameful, and advised me to send in testimonials from Sir Edward Parry and Sir James Ross; which I lost no time in doing, and very strong ones too. Yesterday I saw the new chart in progress at the Hydrographical Office, on which Mr. Walker pointed out to me my own name given to an inlet in Melville Island, by my friend McClintock. On the 28th, meeting Admiral Beaufort in Barrow's office, he said, "So, doctor, you are not made inspector-general yet. You should get Barrow to make me one of the lords, and you would soon get it." Expressed in his goodnatured, friendly way.

On the 3rd of November I had a long talk with Professor Owen, at the College of Surgeons, on the treatment I had met with at the hands of the Admiralty, when I learnt from him that the article in *Fraser*, under the title of "Forlorn Hope," in June last, a review of my "Boat Voyage," was written by Mr. Broderip, the barrister and London magistrate. Whilst with him, my old friend Charles Enderby came in with a magnificent specimen of ambergris, worth at least 5*l.*, to show him.

On the 14th, calling on my good friend John Barrow, at the Admiralty, he told me that he had written to the director-general to know why I was not promoted, and if there was anything against me; and I learnt from him that he received a very stiff reply in return, stating that there was nothing for or against me; that I was a stranger to him, not having seen me for years, except in the streets. Barrow added that he was so indignant and disgusted with the perusal of it that he tore up the note into fragments.

Saturday, February 24th, 1855.—Just as I had left the Admiralty, when near the "Ship Hotel," I most unexpectedly met my old friend Judge Montagu, with his daughter, grown up to a very beautiful girl since I last saw her. They had only returned some three months ago from the Falkland Islands, and were to sail for Sierra Leone on the 23rd of next month, and going from thence on to Melbourne. His son, he told me, was in the navy, and with the Black Sea fleet. He invited me to come and see him at No. 27, Upper Berkeley Street, Portman Square.

Tuesday, March 6th.—At four p.m. I attended Professor Owen's introductory lecture at the College of Surgeons, and at eight p.m. Dr. King's lecture on the Franklin Search, at the Russell Institute in Great Coram Street. After giving a general outline of the search, he drew attention to Lieutenant Pim's and my own share in the search, adding that had my offer when on the spot been accepted, to explore Smith's Sound into the Polar Ocean, the present necessity of a search for Dr. Kane would have been anticipated; and he asked why his own name and mine were not included in the Arctic Council.

On the following evening I was invited by Dr. King to attend the microscopical meeting at Apothecaries' Hall. Met Dr. McWilliam and many other friends there. I learnt from McWilliam that the director-general would not resign his post at the Admiralty until May. On Saturday, the 10th, calling on Dr. King, I was introduced to Earle, the painter of his beautiful picture the "Dog and Moor-hen's Nest." At four p.m., when at Owen's lecture at the college, the vice-president, Mr.

South, asked me to dine with him to-day at Blackheath Park, to meet our old friend Dr. Daniell, before he again departs from us for foreign service. He and I went down together by the 5.30 p.m. train.

Tuesday, 27th.—I walked across Hyde Park to Berkeley Street with a copy of my bound-up "Boat Voyage" for my friend Algernon Montagu. His daughter, Miss Montagu, and his cousin, Mrs. General Hartley, at whose house they were staying, received me in the drawing-room, and soon afterwards the judge himself came in. Whilst he and his daughter were looking over the plates of the narrative, I had some talk with Mrs. Hartley, a clever woman, the authoress, I believe, of several works, if not published, in manuscript, and a linguist possessing great vivacity of manner. She said my name was a very familiar one to her, and was one of the oldest Irish names; that I bore a striking resemblance to a friend of hers in the north of Ireland, and, as my father was a native of Tyrone, she thought I must belong to the same family. She gave me an invitation to come and see her drawings and works some other time.

Monday, May 14th.—I attended Dr. King's lecture at the Marylebone Institution. After passing some severe censures on the bungling of the Board of Admiralty, and the incompetency of the commanders sent out, he said the Franklin Expedition might have been saved had the Admiralty early in the search permitted my carrying out my proposed plan by the Coppermine River, in conjunction with his own down the Great Fish River. He paid me the compliment of saying I possessed the most complete knowledge of the Polar regions, both north and south, fitting me for such an enterprise, and that I might have had the pick of the best men in the service to follow me.

I ought to have noted in its place, the 17th of April

last, the having removed to Pemba Cottage, Adelaide Road, Surbiton, which I furnished, and lived in three years.

Thursday, July 19th.—I attended the First Lord's, Sir Charles Wood's, *levée* at the Admiralty. My interview was of the briefest. He shook hands on my entrance, but little passed beyond bows. I placed my memorial for promotion in his hands, and, as a voucher for the statements therein, having presented him with a bound copy of my "Boat Voyage" to refer to at his leisure, I bowed and retired. On the 30th I received an answer in the negative, as usual.

Wednesday, April 23rd, 1856.—The reward of 10,000*l.* for the discovery of traces of the Franklin expedition being about to be settled I wrote another letter to the Admiralty, accompanied by a memorial of my services, asking not for any portion of the reward, but my promotion to the deputy-inspectorship, and told my friend Barrow that it was now my intention, if possible, to get my case before the House of Commons, if the Admiralty did not acknowledge my services. He said it was quite useless my making any further applications to the Board, as they were not gone into, but that everything referring to my department was left in the hands of the director-general; and on my putting the question to him, as to whether he thought their lordships were prejudiced against me personally, he replied emphatically, "*No;*" had such been the case, *he* must have known it; but that it was not unlikely the late director-general might have left his sting behind him.

Saturday, May 3rd.—I called at the Royal Geographical Society, on the secretary, Dr. Norton Shaw, who gave me a note of introduction to the president, Sir Roderick Murchison, which I delivered to him at the Museum of Geology, Jermyn Street. Sir Roderick promised to write to Admiral Berkeley about my promotion, and also gave me an introduction to Mr. Mackinnon, M.P., of No. 4, Hyde Park Place, advising me to place my testimonials in his hands; and, shaking hands with me, wished me success.

Monday, 5th.—I went to Mr. Mackinnon's, and had an interview with him in his fine picture-gallery. He received me very civilly, shaking hands, and introducing me to a colonel who was with him at the time. He recommended my seeing Mr. Macartney, M.P., who was about bringing forward a motion in the House, and to call upon him again afterwards. I offered him a copy of my "Boat Voyage," which he said he should be happy to accept. He told me that in Captain Maclure's case, had he not succeeded with it Maclure would have been for ever done at the Admiralty. "However," he said, "his was a strong case, and circumstances happened to be very favourable at the moment for carrying it through." He added that he did not like taking up Colonel Chesney's case, which he had been asked to do, lest he might be considered a Don Quixote in these matters. I called at once on Mr. Macartney, at No. 46, Duke Street, St. James's, but was informed that he was at his club, the Carlton, where I had an interview with him. But he told me he never interfered in Arctic matters, being quite unacquainted with the subject, only he had been asked to bring the motion forward in the absence of Sir Thomas Dyke Acland, for Arctic papers to be laid before the House, at the request of Lady Franklin and Miss Cracroft. He told me that Mackinnon carried the House by surprise in Maclure's case, and that he did not himself approve of the claims of officers being brought before the House of Commons.

On the 7th I called again on Mr. Mackinnon, and presented him with a copy of my narrative, which he asked me to put my name in, and on my pointing out to him his own name on the title-page, he asked if it was in my own handwriting, as he said he should treasure up the book. This was followed by a long conversation, in which he told me that he could not bring my claims before the First Lord; nor did he like to take the initiative steps in the House, but should Arctic subjects be again introduced, and I could get Sir Thomas Dyke Acland, who he said was a good man, incidentally to bring forward my name on the occasion, he would himself back it up, all in his power to do; shaking me most cordially by the hand on leaving, adding that he should always be glad to see me at any time. I afterwards called upon my old friend Professor Owen, at the College of Surgeons, and made him acquainted with my proceedings. He promised me that he would see Sir Roderick Murchison in a day or two, and discuss with him as to the best way of bringing my claims forward. On the 10th I called on Mr. Broderip at his chambers in Raymond's Buildings, Gray's Inn, to thank him for his very flattering and friendly notice of my "Boat Voyage," in the very able article he had contributed to *Fraser*, under the title of the "Forlorn Hope," and that I had been for some months past puzzled to know by whom it was written, till our mutual friend, Owen, dispelled the mystery by telling me who it was. He said he never wrote anything more from the heart than the notice of my narrative in *Fraser*, and promised to write to Murchison about me, and refer him to the article "Forlorn Hope" in *Fraser*. I showed him a copy of my letter to the Admiralty and Sir Francis Beaufort, when he said he did not wish to make me too sanguine, but he thought I must get my promotion now, unless red-tapeism stood in the way. He chatted with me in the most friendly way about my night in the snow-drift near Caswall's Tower, with the thermometer at 32° below zero; making me take his easy chair opposite to his friend and relative, to whom he introduced me in the most flattering manner. On the 12th I again called on Broderip, and presented him with a bound copy of my "Boat Voyage," which he expressed himself much pleased with, and said he had written a letter to Murchison, such as he could make public use of; passing some caustic remarks on red-tapeism and its ill effects, pointing out to his relative, who was breakfasting with him, the injustice in my own case, saying that the Admiralty had selected me as an able and competent surgeon, not only to take charge medically of an expedition, but also to command it, thus involving the power of life and death as much as if I had been an executive; and after I had accomplished my object successfully, turn round upon me, and say, "We cannot promote you, because you are a surgeon." Called on Professor Owen at the college afterwards, and gave him a bound copy of my narrative, and showed him my letter to the Admiralty, which he said was capital, nothing could be better; that its businesslike style was in itself a recommendation, and that he would see Murchison some day this week about me.

Next morning, seeing in the *Morning Post* a gazette of the promotion of three deputy-inspectors, for services in the fleet during the war with Russia, I lost no time in seeing the director-general, Sir John Liddell, who has succeeded to that office, about it, and to ask why my own claims, at such a favourable opportunity for acknowledging them, were again ignored; telling him that I had read the promotion in the papers, with the keenest feelings of disappointment and mortification at being again thus passed over. He replied, in his usual bland way, that they were confined to war services, and, moreover, that he knew nothing of it until two days before, when he was sent for to report upon the claims of these three officers the Admiralty had themselves named for promotion, adding that my own was a very hard case, but that his was a very difficult position.

On the 22nd I called on Sir Roderick Murchison, who told me he had not yet written to Admiral Berkeley, as he had promised me, thinking it better to have a personal interview with him, but if I would take a seat he would write me a note to the secretary, Mr. Bernal Osborne, which he read to me, telling me to take it to the Admiralty. It asked for promotion because of my distinguished position as a geographer in the search. I afterwards called on my old friend Dr. Andrew Smith, the director-general of the army medical department, and had about half an hour's conversation with him on the two services in his office, stating the treatment I had met with from the Admiralty. Shaking me heartily by the hand, he wished me every success, adding that he should be most glad at all times to see me, and that the pass-word, by appointment, would be sufficient with his messenger to insure me admittance at all times.

Monday, 26th.—I again saw my own director-general, Sir John Liddell, and told him that all my friends believed my promotion rested with him. He replied, "I hope they will not get that impression, as it is not the case. I strongly recommended you last time at the Admiralty; but their lordships appear more disposed to reward present than past services." I replied mine were comparatively recent, and that I had been unceasing in urging my claims, and that the friends who had so kindly moved in my behalf considered my boat voyage *alone*, independent of my other services, both professional and to geography and science, ought to claim for me my promotion, and that I looked to him, as the head of my department, to bring forward my claims on purely surgical acquirements, which had gained for me the "honorary fellowship" of my college. He said that he had received a letter from the vice-president of the college, Mr. South, about me, and very strongly recommending me. Mr. Barrow and everybody else, he said, did the same thing, and what everybody said must be true, adding, "And I myself entertained the same opinion from the first; and told Mr. Barrow that you should make your application for promotion to the Admiralty, and I would do as I have always done, support it; but that Mr. Bernal Osborne's writing to himself was useless, as he had no power beyond recommending." From Sir John's manner in shaking hands with me as I was leaving, my interview, so far as I could judge, had made a favourable impression on him, as my friendly supporters undoubtedly had by his own admission. On the 28th, calling again upon Sir Roderick Murchison, he told me he had done all that he could; but they were a heartless set at the Admiralty. He seemed much discouraged with the failure of his efforts to serve me, which is not at all surprising.

Thursday, June 26th.—I again attended Sir Charles Wood's *levée*. On handing him my memorial, with Sir Francis Beaufort's letter, he read them both, then referring to the "Navy List" for my seniority, remarked, "Are you the senior surgeon on the list?" I said, "Yes, for active service." He replied, "Quite right; I will see Sir John Liddell about you;" and so ended our interview, like so many others, anything but satisfactorily.

Wednesday, December 10th.—I attended Dr. King's meeting at the Russell Institute, and took my seat on the platform about eight p.m. Dr. King opened the meeting with a long speech, followed by one from Lieutenant Pim; after him Mr. George A. Findlay spoke; and then I myself rose, and addressed a few words to the assemblage. On resuming my seat, Dr. King introduced me by name to the audience, paying me the compliment of being a sound and cautious geographer, whose opinions should carry great weight with them in the search. After myself, Mr. John Brown rose, and pointed out on the chart his own views in opposition to mine, denying that Franklin ever went north of Barrow's Strait. Brown was followed by Bertholdt Seamann, the naturalist, in a long speech. There was a good attendance, and a fair sprinkling of ladies. I left whilst Dr. King was speaking, at ten p.m., to secure my train home.

On Saturday, the 20th, I went to the last of King's lectures at the Marylebone Institute, at eight p.m. Met King at the door going in. He made a long speech, followed by Pim; and next by myself, stating that the records alluded to by King could not be too much dwelt upon; and I ended by adding that I, for one, would never cease to agitate in favour of further search so long as there was the shadow of a hope, and that whenever the "Blue Peter" was hoisted, would be the first to serve, and was warmly cheered. Afterwards a gentleman near the platform made some remarks about Arctic men not entering into the discussion, when Dr. King rose and said that Mr. McCormick, who had just spoken, was the most experienced amongst Arctic officers, having been at both the Poles.

January 8th, 1857.—I laid my last plan of search, by Bellot's Strait to King William's Land, before the Admiralty, offering myself to conduct it, and accompanied by a note to Barrow. I also put a copy of it into the hands of Sir Roderick Murchison, to lay before the Geographical Society. On the 3rd of February, on going to the society to see Dr. Shaw, the secretary, he told me that Sir Roderick had given him my plan of search to record, which I found had been done in due form.

April 1st.—Calling on Mr. Weir, the editor of the *Daily News*, to ask him to review Dr. Armstrong's narrative of his voyage, I found him confined to the couch in his office, very ill, poor fellow, and indeed in a hopeless condition; when he replied, "Will you review it for me? I shall esteem it a favour." And on the 20th

I went to his office again with the review I had written; and after he had heard it, I asked him if he saw anything to alter. He replied, not a word; that he liked it much. On the 23rd it appeared in print, filling about three columns of the paper.

May 12th.—A beautiful young barn owl I had scarcely had a year, which, from his extraordinary intelligence and affectionate disposition, had so attached me to him as to become a great pet, died in my hand about midnight. He had been very ill throughout the day, and died in convulsions, and all my efforts to preserve the life of the poor "Major," as I had named him, proved unavailing.

June 18th.—Being the Waterloo anniversary I attended her Majesty's *levée*, wearing the "Arctic Medal" I had lately received for the first time, some eighteen years having passed away since I was last at court, being then my *first* presentation.

On July 2nd I attended the First Lord's *levée* at the Admiralty at four p.m. Sir Charles Wood rose from his seat, offering his hand, and desiring me to be seated. As this was our third interview, I reminded him of his promise last year to see Sir John Liddell, and go into my claims; adding, I had hoped now to have had the gratification of thanking him for my promotion, instead of having to trouble him again with a repetition of my services. And as he took up the "Navy List" to refer to my name, I remarked that he would find it, heading the "active list;" at the same time calling his attention to the "deputy-inspectors' list," that out of the fifteen officers comprising it, fourteen were my juniors as surgeons; and that even in the "inspectors' list," four out of the six had been junior to me as surgeons; that I had been engaged in three Polar voyages, served three times in the West Indies; had been thirty-four years in the service, thirty of which as surgeon; had been a volunteer both for the Baltic and Black Sea fleets, at the commencement of the war, on my return from the Arctic regions. I then bowed and retired, leaving his lordship making notes of what I had said.

January 30th, 1858.—I attended the Queen's drawing room at St. James's Palace, the first I had ever been at. On Wednesday, the 7th of July, I was elected a member of the Junior United Service Club without a single "black-ball" in the box, for there has been some "black-balling" of late of the candidates amongst the militia. On the 19th of February, 1859, I proposed my old friend Dr. Daniell, now absent on foreign service in the army, seconded by Dr. Dobie, a brother officer of mine. On the 25th I went to the House of Commons; had a seat in the Strangers' Gallery. There was a full house, to hear Lord Palmerston's motion on the state of Europe; after which, at 6.30 p.m., Sir John Pakington rose, and sat down again at nine p.m. In giving the Naval Estimates, he paid the naval surgeons a very high compliment, and pledged himself to grant them the "warrant" in spirit and in substance. I left the House sitting at 1.30 a.m.

On Thursday, the 24th of March, I removed from St. John's Wood to Wimbledon, having taken a newly-built villa but just finished, in a new road called Ridgway Place.

On the 20th of May, 1859, I was at last promoted to the deputy-inspectorship on the "Active List," on the very eve of the "new warrant" coming out. After so many repeated disappointments and hopes deferred, till it was literally "hoping against hope," I was in the "eleventh hour," when least expecting it, one morning surprised by reading in my morning paper, the *Standard*, at breakfast-time, the talismanic words, "promoted to deputy-inspector-general," attached to my name. Such was the medium through which I first became acquainted with my promotion. On meeting my old brother officer, Dr. McWilliam, of Niger Expedition celebrity, he warmly congratulated me with a hearty shake of the hand, and said my promotion had given great satisfaction to the Service, and that I owed it to no one but myself and my own determined energy, condemning the late director-general's conduct towards me in no measured terms of censure and contempt. Upon my calling upon the present director-general, Sir John Liddell, to ask him to whom I owed this somewhat tardy acknowledgment of my past services, he replied at once, "Entirely to Sir John Pakington;" and, shaking hands with me, said, "Are you satisfied now, and have you got what you want?" I replied, "I must be so; but had the present step been conferred on me at the time the services I had rendered claimed it for me, I should now have been at the top of the 'inspectors' list,' instead of at the bottom of the deputy-inspectors, nearly all the inspectors on the list having been my juniors in the service, and with no very unusual or extraordinary claims." To this he remarked that many of these promotions were "accidental."

On the 25th of June I attended her Majesty's *levée* at St. James's Palace on my promotion, having previously waited upon the Duke of Somerset, the new First Lord, at his residence in Wimbledon, he having not yet entered upon his official duties. I delivered a letter of introduction to him from my old friend Sir Roderick Murchison, when he kindly signed my ticket for the *levée*, being the only one, I believe, signed by his Grace, with the exception of Admiral Sir Michael Seymour, during the interregnum occasioned by the change of ministry. I wore at the *levée*, for the first time, the new deputy-inspector-general's uniform, that of a post-captain, with the exception of having a crown with a star, instead of the anchor, &c., on the epaulettes, and the coat single-breasted.

February 7th, 1860.—Dr. Daniell, now absent in the

China war, whom I proposed on the 19th of February last, was this day elected a member of our club, the Junior United Service.

On Thursday, the 5th of April, I attended the First Lord's *levée*. I was the first on the list, and on entering the room the Duke of Somerset rose from his seat, shook hands, and requested me to be seated. I presented his Grace with a copy of my "Boat Voyage" in search of Franklin, adding that since last year, when through his kind consideration I was enabled during the ministerial interregnum to attend the Queen's *levée* on my promotion, presented by himself, the fate of the Franklin Expedition had been discovered, the plan of search for which had originated with myself, I having drawn it up and had it laid before the Geographical Society by its president, Sir Roderick Murchison, and also had laid it before the late Board of Admiralty, with an offer on my own part to conduct it—a few copies of which I had now bound up with the narrative of my boat expedition for private circulation amongst a few friends interested in the search. The plan itself had been ably carried out to the very letter most successfully by my friend, Captain McClintock, R.N., under the patronage of Lady Franklin, in a ship of her own. The duke accepted the book, thanked me, and on opening it remarked, on looking over the plates, "It is very nicely got up, and an interesting subject." On my rising to go, his Grace got up from his own chair, and again shook me by the hand.

June 6th.—I called on the Baron Marochetti, when he showed me the bust of Sir John Pakington, and the statuettes in terra-cotta and plaster-of-Paris, in his studio; I selected one of the terra-cottas for myself as a subscriber to the bust.

On November 28th I removed my mother and sisters from St. John's Wood to Fir Villa, Surbiton, which I had taken for them.

On February 7th, 1861, my poor dear old friend John Brown, of Scalesby Lodge, died of congestion of the lungs; he was an old and most valuable Fellow of the Geographical Society, and the author of the best work on the Franklin search. I saw him during his illness and on the day before his death, and attended his funeral at the Highgate Cemetery, where he was buried. John Barrow, John Arrowsmith, Dr. Norton Shaw, and both the Findlays, father and son, were at the funeral. A better man never lived, generous and warm-hearted, and a staunch friend.

March 11th.—When dining with my landlord, Mr. Mackeril, a solicitor, of No. 25, Abingdon Street, at his residence in the Wandsworth Road, I expressed a desire to become owner of the villa I had resided in for the last two years, and although he said he did not wish to part with any of his property in this road, most of the houses belonging to him, he met my wishes, and the house became my own by purchase on the 21st of May.

April 7th.—I made an excursion to Boulogne, going by the *Panther*, on the Sunday, and returning by the *Maude*, on Tuesday, the 9th, to Folkestone; stopped at the "Hôtel de l'Europe," Boulogne.

June 15th.—Accompanied my old friend, Dr. Daniell, to the British Museum, where he gave me a specimen of the New Tartary oak he discovered, and had named after me.

August 9th.—My poor mother breathed her last at Surbiton, and I alone attended her funeral on the 15th at Kensal Green.

May 8th, 1862.—I attended the Duke of Somerset's *levée* at the Admiralty; it was a very hurried affair, I was the eighth on his Grace's list, and the interviews with that number did not occupy half an hour. He seemed to have made the audience of the briefest with all, from what I heard. On my asking him for my promotion to the inspectorship specially for special claims, all he said was, "Very well."

On Friday, July 11th I entered the great exhibition for the first time, having taken a season ticket.

On February 19th, 1863, I again made my appearance at the Admiralty, at the duke's *levée*. I reminded him of my special claims to the inspectorship for Polar service, and also stated that having annually sent in my offer to serve, and now seeing no hope of being permitted to complete the required five years for the next step, at my age, I hoped this would be taken into consideration. His Grace replied, "I will make a note of your claims."

On the 25th I made my first appearance at the Prince of Wales' *levée*, at St. James's Palace.

May 5th, 1864.—I once more paid what has now become my annual visit to the Duke of Somerset's *levée* at the Admiralty. I happened to be nearly the last on the list, and, walking up to the table, placed my memoir in his Grace's hands, then bowed and retired.

On the 30th of March, 1865, I again presented myself at the Duke of Somerset's *levée*, at the risk of being considered importunate; but when without powerful friends or political influence, it will not do to "lay on one's oars" in the struggle for advancement in this life. And as the duke had given as his only reason for not promoting me my not having served the regulation five years, on his requesting me to be seated, I referred him to the Order in Council of the 26th of November, 1858, which provides for the special promotion to inspectorial rank for meritorious services, without having served the usual time. He then took up the Navy List, and from it made some memoranda in his note-book; saying at the same time that he should be most glad to meet my wishes, were he not so pressed by others with long services that he could not say he would depart from the rules of the service. I said I was fully aware of the difficult position he was placed in, and consequently should not so urgently have pressed my claims at the present moment, had not my position on the active list become so critical a one that any future retirement scheme might shelve me at once, without any recognition whatever of past claims. Feeling, as I did, active and as fit as ever for service, my sole wish was to avert my compulsory retirement; and from this promotion to the inspectorship could alone secure me, by giving me five years longer on the active list. The duke replied that he quite understood me, and patiently heard all I had to say, during a longer interview than ordinary, some ten or fifteen minutes.

The realization of my anticipations and apprehensions of being shelved occurred sooner indeed than might have been expected. On the 14th of the following July I was summoned to appear at Somerset House to be surveyed, notwithstanding I had already answered a letter, stating my fitness and readiness to serve whenever called upon. And now I could only repeat this *vivâ voce.* When the present director-general, who succeeded Sir John Liddell, asked me if I was willing to go to Hong-Kong, my answer was, "Yes, I am, or anywhere else in any part of the world, or with the Channel Fleet, or, for that matter, to the North Pole again if required, although I did not get my last step in rank until sixteen years after it had been earned in the voyage to the Antarctic Seas, and since then, although I had annually offered to serve, no opportunity was afforded me for serving my five years for the inspectorship, and I had even been removed from the *William and Mary* yacht at the expiration of three years without any equivalent compensation being offered me; yet the yacht was considered a life appointment when I joined her." He again asked me if I was ready to go to any part of the world as well as to Hong-Kong. Certainly, I said, rather than be shelved, and asked him if that was the object of the survey held on me. He said he could not answer that question, as they—meaning his assistant in the office, who was leaning against the mantelpiece—did not know what was done at the other house in Whitehall (Admiralty). After this there could be no doubt whatever of the intention of the authorities to shelve me. Consequently no time was to be lost in moving what friends I had to checkmate them in this, if possible. I at once called upon my old friend, Professor Owen, at Sheen Lodge, Richmond Park, on the following day but one. There I met, and was introduced to Mr., Mrs., and Miss Roebuck, who were making a morning call. As the professor and Mrs. Owen insisted on my staying to dinner, it afforded me a favourable opportunity of making him acquainted with the shabby treatment I had met with at Somerset House, when he kindly offered to see General Sabine about me, and eventually gave me a note to him, asking him, as President of the Royal Society, and himself about the last of the old Arctic men, to lay my memorial before the First Lord. With this I called at Ashley Place, Pimlico, when General Sabine, after reading it, remarked that it was a very strong case, but the difficulty lay in the way of bringing it before the First Lord. He thought it would not do to put it into his hands himself, and wished first to see the secretary of the Royal Society and get a minute made of it there, as he wished to write a strong letter, and would forward it to my club before he himself left town to-morrow for a six weeks' tour in Wales. He remarked on the Order in Council for special promotion, that mine was, in reality, a case in point to meet it; and added that he had had much to do with the Antarctic expedition, and could answer for me himself. I was altogether received very kindly. Mrs. Sabine rose from her seat and shook hands on my departure, after nearly an hour's visit, and the general himself accompanied me to the door, giving me a hearty shake of the hand. On the 26th I placed my memorial, with Owen's and Sabine's letters, in Captain Hall's hands, at the Admiralty.

July 29th.—I received a letter from the Admiralty, stating that I was placed on the retired list, and on the 11th of August a refusal to place my name on the retired list of inspectors.

November 13th.—I placed a copy of my suggestions for the improvement of the medical service of the navy in the hands of the commissioners now sitting on that subject.

On the 16th I called on Mr. Mackeril, No. 25, Abingdon Street, and purchased of him No. 2 Villa, opposite to my own in Ridgway Place. On the 9th of March I attended the last *levée* of the Duke of Somerset, and during some five minutes I had to explain everything in reference to my late compulsory retirement. His Grace patiently and quietly listened to me, offering me a seat, and on my placing in his hands a copy of my narrative, to which I had recently added my printed testimonials, he remarked, "You gave me a copy some time ago." So that it appears his Grace does not forget things at all events, for seven years have elapsed since then. I observed that it was very discouraging to find two brother officers, who had decidedly refused the appointment to Hong-Kong, given the honorary rank of inspector because they chanced to have completed the time for it, whilst my own name was ignored because I fell somewhat short of the regulation time, although a willing volunteer for Hong-Kong. I said I hoped at least I had some claim to the Greenwich Hospital pension, the only thing left me now.

On the 26th of September, the Greenwich Hospital

Pension becoming vacant, I applied to the Admiralty for it, but with no better success than when I asked the Duke of Somerset for the first one given at his last *levée*. It was given away on the 12th of October. On the 16th of May, 1867, I attended Mr. Corry's (the new First Lord) *levée*. He rose from his seat, shook hands, and asked me to be seated. I asked him at once to place my name on the retired list of inspectors, and pointed out to him that the only Greenwich Hospital pension for deputy-inspectors had both times been conferred upon honorary inspectors, and that I had been most unjustly treated by the late Board. He replied he would see the director-general on the subject. I said that would be of no use, as I believed the injustice originated in that quarter. I offered him a copy of my narrative, in which, if he would glance at the appendix, he would find my long Polar services therein stated. "Polar services," he re-echoed, as if somewhat surprised, strange as it may be, and thanking me, said he should be glad to read the whole. He then added he would see the lord who presided over our department, about me, who, he believed, was Sir John Hay. On my rising to go he got up from his chair and again shook hands with me.

On the 30th of September I left by the train from Charing Cross, accompanied by a friend, for Paris, to see the great Paris Exhibition, viâ Folkestone and Boulogne; stayed at the "Hotel Brighton." Entered the exhibition on the 1st of October, and after devoting four days to exploring it and the magnificent grounds around it, returned on the 5th, viâ Calais. After my return I was confined to my room by a dangerous illness, and could not leave the house until the 13th of January, 1868, when I went over the way to see my sisters, being the first time I had been out of the door for a month past.

On Tuesday, the 5th of May, I attended my last

Admiralty *levée*. The interview with Mr. Corry this time did not exceed five minutes, and the whole of the twenty-five officers about an hour. Mr. Corry, when the messenger announced my name, was standing by the table, bending over some papers. He moved towards his chair, extending his hands, and motioned me to a seat opposite to himself. I reminded him of his promise to see Sir John Hay about me. He replied, he did not know whether he had done so or not, and making a note in his memorandum-book, very emphatically said, "I'll make a point of seeing him now." On my stating that I was the only one retired of all the Polar men of every grade whose services remained wholly unrecognized, his memory would appear to have failed him so much since our last interview that he actually asked me if I was not still on the active list; when I placed in his hands a copy of my last year's memorial, bowed and retired.

On the 3rd of September, 1876, the Greenwich Hospital Pension once more becoming vacant, I again applied for it, and on the 11th of October it was at last conferred upon me, thus adding to my annual income 80*l.* a year. However desirable this boon may have been to me in a pecuniary sense, had the promotion to the rank of inspector-general, the highest step in our department of the service, accompanied it, such an honorary distinction would have enhanced the value of the pecuniary reward, inasmuch as the executive class of officers on retirement have a step in rank conferred upon them without any claims to special service being required of them. So that in my own case, services so much out of the ordinary routine, and which I may, perhaps, without presumption, term unique in character, as I think these memoirs will amply testify, might claim for their author the highest step available without any additional expense to the country.

Therefore, on the return of the recent expedition, under the command of Captain Sir George Nares, from the attempt to reach the North Pole, the very liberal promotion of the officers engaged in it encouraged me, on the 6th of the following December, to make a final effort to reach the top of the tree, the ambition of all, by addressing a memorial to Mr. Ward Hunt, then First Lord of the Admiralty, asking him, in consideration of my long and varied services, to place my name on the list of inspectors-general for honorary rank only, forwarding him at the same time a private copy of the narrative of my boat voyage up the Wellington Channel in search of Sir John Franklin. This, my final effort, was also unsuccessful.

In resuming my autobiography from the year 1876, there have been few incidents worth recording; but to preserve an unbroken chain of events, I will bring them down to the present time. Towards the close of that year, death removed an old friend and neighbour of mine, Sir Alexander Campbell, Bart., of Barcaldine, N.B. His widow, the daughter of the late Admiral Collier, was a niece of the late Admiral Sir Francis Collier, under whose command I had formerly served in her Majesty's yacht, *William and Mary*, at the time he was superintendent of Woolwich Dockyard. Lady Campbell, with her two daughters and her eldest son, the present Sir Duncan Campbell, Bart., are still my neighbours, and their continued friendly visits tend to relieve the quiet seclusion in which I live; together with an occasional visit or letter from my old and esteemed friend, John Barrow, whose name so frequently appears in these pages, and who, since his retirement from the Admiralty, has been one of the most zealous and enthusiastic supporters of the volunteer force, from which he has now retired with the rank of colonel, and who is as ardent as ever in the cause of Arctic discovery. My young friend, D'Eyncourt

Chamberlain, the son of a retired paymaster-in-chief of the navy, also visits me whenever he can spare time from his medical studies at University College, where he is preparing to enter the navy as a surgeon, and whose mother was the daughter of very old friends of my youth; and last, though not least, the occasional correspondence with the three sons of my late lamented old friend, John Brown, who was a distinguished fellow of the Royal Geographical Society, and the author of the best work on the Franklin Search.

I have somewhat entered into minor details here to give the reader an insight into my domestic life, whilst employed in preparing my work for the press; and some description of the very remarkable pet white Aylesbury duck, the lithograph of which comes in at the end of these memoirs, may claim a place here, from her having assumed in old age the curled tail-feathers of the drake, a unique, I believe, and interesting fact for the ornithologist.

The "Duchess," as she has been called, is a *lusus naturæ* and *rara avis* amongst ducks, which I have never before met with, nor, as far as I can ascertain, have the authorities of the Ornithological Departments of the South Kensington Museum of Natural History.

Her history is as follows: she was presented to me when a little duckling, clothed in yellow down, only ten days old, in the month of May, 1870, and is consequently in her fourteenth year, being the only survivor of a brood of eggs—and from the finest white Aylesbury breed—hatched under a hen. From the first she showed marked features of originality, intelligence, and observation, and a wonderful memory, an affectionate nature, and sociable disposition. She sleeps in a wicker-basket made especially for her, and has now, in her old age, forsaken the garden and the duck-pond, living entirely within doors.

Her favourite place is upon a cushion, made from her own feathers, on the hearth-rug before the sitting-room fire; and on a winter's evening, after the lamp has been lighted, on my rising from my chair, where I may have been writing or reading, for a turn up and down the room, on my saying to her, "Ducky, come and have a quarter-deck walk," she will at once answer me in her own peculiar tone of voice, leave her cushion, and waddle up and down the room, close at my heels, and at each turn pull at my slipper with her beak, as much as to say, "Go on," till she is tired out, when she will suddenly stop, look up at me with an expression that would say, "There, that is enough for once." She understands all I say to her, as a dog would do, being equally sagacious and companionable. Her control over her voice, in modulating its tones to express her feelings, is most extraordinary, from her customary soft and pleasing notes to a sharp rebuke, when displeased, that indeed, from long observation of her and her ways, I have no difficulty in interpreting all her wants and wishes.

She commenced laying before the end of her first year, and so prolific has she been, that during the first year, she laid about 120 eggs, and not a less number in successive seasons until her tenth year, when, showing a strong desire to sit, I procured for her, from Baily's of Mount Street, four eggs, of ornamental ducks, but after sitting on them with the greatest assiduity for the usual period, they all turned out bad. On my acquainting Baily with the disappointment she had been subjected to, he sent me six ducklings in a basket, hatched in the morning of the day on which I received them at nine p.m., and having at once placed them under her in her basket, she readily took to them and reared them as her own, three of them at least, a mallard and duck of the wild breed, and a white "*call-duck*," the other three died in the first week, probably from her having laid upon them. The wild duck, an intelligent, very pretty creature, died last autumn of a chronic disease of some standing, only five years old.

The two pairs of curled tail-feathers which normally distinguish the drake from the duck, in the white Aylesbury breed, made their first appearance in the "Duchess" during her autumnal moult in 1880, in her tenth year, after she had ceased laying, since which she has shed, and reproduced them twice a year in her annual moults; but they were most perfect in form and size in the first two years, since which they have decreased in size and form, and become more feeble in appearance each succeeding moult. Her last eggs laid were somewhat abnormal in form, elongated, and kidney-shaped. This singular change in age to the plumage of the male is evidently due to some change in the condition of the ovaries.

She is as companionable and sociable as the dog, and will follow me in my walks anywhere. If I ask her to shake hands with me, she will at once place her beak in the palm of my hand, accompanied by a peculiarly pleased chuckle; is as lively and playful as ever, though somewhat tottering and unsteady in her gait, and does not now leave her basket in the corner of the room until past midday. Her appetite is good, but requires constant change, variety of food, and frequent baths.

At night her supper is invariably bread and milk, but during the day watercresses, wheat, boiled chestnuts, macaroni, tea and sugar, of which she is very fond, and many other things she has been taught to take by the members of my household, with all of whom she is such a pet and favourite—there is nothing they will not do for her.

When Mr. Edwin Wilson came down to Wimbledon to take her likeness for the lithograph—which he ultimately finished at the Kensington Museum—

in which he has so happily caught the intelligent glance of her expressive eye, he was much astonished that whilst her likeness was being taken, she stood on the table between three and four hours, as if she thoroughly understood what was going on, by the calm, intelligent glances with which she eyed him, and he remarked that she was a better sitter than many of us would have been, so self-possessed was she throughout the whole of the time.

I hope that the somewhat lengthy description I have been led into, by the animus which has impelled me to bring under the ornithologist's notice especially so unique a specimen of a remarkable aberration from nature's laws, will not tire the reader's patience, more particularly as I cannot dismiss the subject without calling attention to the little bird with uplifted wings on the cushion beside the "Duchess,"—a hen sparrow, called "Polly," as unique in some respects, which I had for above seven years. I picked her up in the garden an unfledged nestling,—another example of what may be accomplished by kindness and training. This companionable, affectionate little creature, although with the full use of her wings, and the whole run of the house, with windows and doors necessarily open, never once attempted to avail herself of the many opportunities she had of making her escape, tempting as must have been the attraction of the society of her race in the garden, more especially during the pairing season. Yet she laid several eggs in a nest I placed for her over the staircase window. Poor thing! she was destined to meet with a sad end, being crushed by the closing of a door, at the moment she was passing from one room to another, in January, 1869. She always took her meals on the table with me, and would perch on my head or shoulder when reading or writing, and sit and plume herself much in the way as the sparrow, described in Professor Wilson's—Christopher North—life by his daughter, which, she tells us, would perch on the professor's shoulder whilst reading or writing in his study, and pull his grey hairs, and appears to have been his constant companion in his studies.

Whilst correcting these proof-sheets, I accidentally came across a quotation in the translation of a French work in my library, stating that Yarrell had met with two instances of curling feathers in the tail of the wild duck, but no mention is made by him of the Aylesbury duck.

As I was bringing this Autobiography to a close I wrote to my old friend Sir Richard Owen, to congratulate him on his being recently created a K.C.B. I received a reply in so friendly a spirit, leaving me to my own discretion as to printing it, that I venture to introduce the letter as a conclusion.

SHEEN LODGE, RICHMOND PARK,
13*th January*, 1884.

DEAR MCCORMICK,—Few congratulations have been more gratefully received than yours, and not any of them gave me such pleasure as your announcement of the completion of your long-looked-for, valuable, and most interesting work. A welcome record that will be of those Arctic explorations, which will ever brightly adorn the annals of the British Empire.

I wish I was as far advanced in my big book on British Fossil Reptiles, which will be in four quartos, two of text, two of plates.

May I live to see it out, and may yours be brightened by recognitions for years to come of your bold and exemplary devotion to Arctic geography. Again thanking you,

Believe me, ever truly yours,
RICHARD OWEN.

Edwin Wilson, del. Vincent Brooks, Day & Son, Lith.

A remarkable White Aylesbury Duck, assuming in age the curled-tail feathers of the Drake, and a pet hen Sparrow no less unique.

TO FACE END OF AUTOBIOGRAPHY—VOL. II.

LETTER TO THE RIGHT HON. GEORGE WARD HUNT, M.P., FIRST LORD OF THE ADMIRALTY, ETC.

Ridgway Place, Wimbledon,
December 6th, 1876.

MY LORD,—1. From the deep interest your Lordship has evinced in Polar Discovery, and from the very liberal promotion of the officers returned from the recent attempt to reach the North Pole, under your Lordship's administration ;

2. I am encouraged to venture to hope that my own humble services when belonging to H.M.S. *Hecla*, in the late Sir Edward Parry's attempt to reach that Pole, now some fifty years ago, may at such a favourable moment as the present, obtain for me the promotion I have so long been ambitious of obtaining—the elevation to the highest rank in the department of the service in which I have spent my best days. Such a reward, in my declining years, and still faster declining health, would, I need scarcely say, be appreciated by me as the greatest boon that your Lordship could bestow on one, all but the last surviving officer of an expedition, which, if it failed in reaching the Pole under so good and able a commander as the late Sir Edward Parry, was simply owing to the wrong season of the year having been selected for boating and sledge operations, with a strong current drifting them back to the south again.

3. Nevertheless, my own conviction is that the Spitzbergen Sea will be found the high road to the North Pole, which will be attained by that route sooner or later. The Gulf Stream, the comparatively open sea at intervals, with a current from the Pole, and many other circumstances all point to this. Whilst the heavy accumulated Polar pack, resting, as it does, on the northern shores of Greenland, and extending across Smith's

Sound, Grant's Land, and the Wellington Channel, to the Parry Islands, remains impenetrable.

4. I trust, my Lord, that your Lordship will exonerate me from the charge of presumption in thus digressing, which I do most respectfully, to venture an opinion of my own as to the causes and in defence of the unavoidable failure of one whose name I so revere and esteem as that of my old commander the late Sir Edward Parry, foremost of Arctic explorers. The only excuse I have to offer your Lordship is the circumstances of the moment, and my own devotion to the Polar discovery from my earliest youth ; and to the executive line of the service for which I had been originally intended by my father, whose untimely loss as an officer of H.M.S. *Defence*, of seventy-four guns, wrecked on the coast of Jutland, prevented from being carried out.

5. I have served in three Polar expeditions, the one in the attempt to reach the South Pole, in which three years and more were spent in buffeting with the vast packs of ice in the Antarctic Seas, which was rendered ever memorable by the discovery of a southern continent. Yet I was the only officer (eligible) left unpromoted. Notwithstanding which I subsequently commanded an open-boat expedition up the Wellington Channel in search of my old ships the *Erebus* and *Terror*, under the command of the ill-fated Sir John Franklin, and I never lost a man by sickness, or had a true case of scurvy in all these voyages, extending altogether over the long period of some seven years, with intervening services in the Tropics, embracing three stations in the West Indies.

6. I am, too, the oldest officer of inspectorial rank, and obtained the full Surgeon's rank before any of the Inspectors or Deputy-Inspectors on the Navy List. I may also state that, in addition to my ordinary medical duties, in all these voyages I voluntarily performed those of Naturalist and Geologist. My tardy advancement to the Deputy-Inspectorship when I was nearly sixty years of age, proved an insuperable barrier in the way of my serving the time required for the Inspectorship. Having had the honour of receiving my first step in rank to Surgeon from our late naval king, William IV., when Lord High Admiral, it would be to me, towards the close of life, a most gratifying climax to a long Polar career, should your Lordship and my

Lords Commissioners of the Admiralty, who have so recently most considerately awarded me the vacant "Greenwich Hospital Pension," be pleased to confer on me the final step in rank in my profession as a "special" reward.

I have the honour to be, my Lord,
Your Lordship's humble and obedient Servant,
R. McCormick, R.N.
Deputy Inspector-General of Hospitals and Fleets, Retired.

The Right Hon. George Ward Hunt, M.P.,
First Lord of the Admiralty, &c., &c., &c.

REPLY.

Admiralty, 16th December, 1876.

SIR,—In reply to your letter of the 6th inst., praying that in consideration of your former Arctic services under Sir Edward Parry, you may be granted the honorary rank of Inspector-General, I am commanded by my Lords Commissioners of the Admiralty to acquaint you that they regret they cannot comply with your request; the fact of your never having served in your present rank, renders you ineligible for advancement to a higher grade.

I am, Sir,

Your obedient Servant,

(Signed) ROBERT HALL.

R. McCormick, Esq.,
Retired Deputy Inspector-General, R.N.
Ridgway Place, Wimbledon.

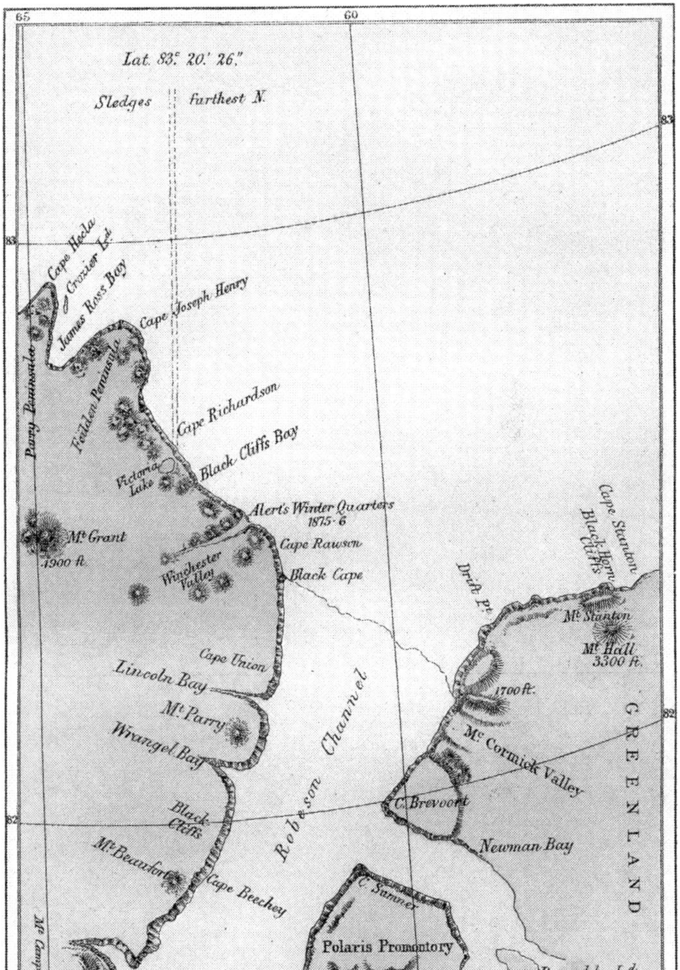

R. Mc Cormick, R.N. Vincent Brooks, Day & Son, Lith.

MAP OF SMITH'S CHANNEL &c.

APPENDIX

OF

PLANS OF SEARCH, ADMIRALTY CORRESPONDENCE, AND TESTIMONIALS.

APPENDIX

OF

PLANS OF SEARCH, ADMIRALTY CORRESPONDENCE, AND TESTIMONIALS.

No. I.

OUTLINE of a PLAN of an Overland Journey to the Polar Sea, by the Way of the Coppermine River, in Search of Sir JOHN FRANKLIN'S Expedition, 1847.

IF Sir John Franklin, guided by his instructions, has passed through Barrow Straits, and shaped a south-westerly course, from the meridian of Cape Walker, with the intention of gaining the northern coast of the continent of America, and so passing through the Dolphin and Union Straits along the shore of that continent, to Behring's Straits;

His greatest risk of detention by the ice throughout this course would be found between the parallels of 74° and 69° north latitude, and the meridians of 100° and 110° west longitude, or, in other words, that portion of the North-west passage which yet remains unexplored occupying the space between the western coast of Boothia on the one side, and the island or islands forming Banks and Victoria Lands on the other.

Should the *Erebus* and *Terror* have been beset in the heavy drift ice, or wrecked amongst it and the broken land, which in all probability exists there, whilst contending with the prevalent westerly winds in this quarter;

The Coppermine River would decidedly offer the most direct route and nearest approach to that portion of the Polar Sea, and, after crossing Coronation Gulf, the average breadth of the strait between the Continent and Victoria Land is only about twenty-two miles.

From this point a careful search should be commenced in the direction of Banks Land; the intervening space between it and Victoria Land, occupying about five degrees, or little more than 300 miles, could, I think, be accomplished in one season, and a retreat to winter quarters effected before the winter set in. As the ice in the Coppermine River breaks up in June, the searching party ought to reach the sea by the beginning of August, which would leave two of the best months of the year for exploring the Polar Sea, viz. August and September.

As it would be highly desirable that every available day, to the latest period of the season, should be devoted to the search, I should propose wintering on the coast in the vicinity of the mouth of the Coppermine River, which would also afford a favourable position from which to recommence the search in the following spring, should the first season prove unsuccessful.

Of course the object of such an Expedition as I have proposed is not with the view of taking supplies to such a numerous party as Sir John Franklin has under his command ; but to find out his position, and acquaint him where a depôt of provisions would be stored up for himself and crew at my proposed winter quarters, where a party should be left to build a house, establish a fishery, and hunt for game, during the absence of the searching party.

To carry out this plan efficiently, the Hudson's Bay Company should be requested to lend their powerful co-operation in furnishing guides, supplies of pemmican, &c., for the party on their route and at winter quarters. Without entering into details here, I may observe that I should consider one boat, combining the necessary requisites in her construction to fit her for either the river navigation or that of the shores of the Polar Sea, would be quite sufficient, with a crew one half sailors, and the other half Canadian boatmen ; the latter to be engaged at Montreal, for which place I would propose leaving England in the month of February.

Should such an Expedition even fail in its main object, the discovery of the position of the missing ships and their crews, the long-sought-for Polar Passage might be accomplished.

(Signed) R. McCormick, R. N.

Woolwich, 1847.

No. 2.

Outline of a Plan of a Boat Expedition in Search of Sir John Franklin's Expedition.

Her Majesty's ship *North Star*, recently commissioned for the purpose of taking out an additional supply of provisions to Lancaster Sound, for the use of the arctic ships now absent, offers so favourable an opportunity for making another effort to ascertain the fate of Sir John Franklin's Expedition, and that, too, without in any way impeding the particular service in which the *North Star* is to be employed, or even involving the necessity of that ship's wintering in the ice.

I feel it my duty, as an officer who has been employed in former Expeditions, and devoted many years past to the subject of Polar discovery, to suggest that Jones and Smith Sounds, at the head of Baffin Bay, should be carefully examined by a Boat Expedition; but more especially the former, it being the first opening north of the entrance to Lancaster Sound.

These openings to the Polar Sea, although most important ones, still remain unexplored, not coming within the sphere of search of any of the Expeditions at present employed in those seas. That they are important ones, I need only quote the opinion entertained by Colonel Sabine, one of the best authorities on this subject, who states in a letter to the Admiralty, that "it was Sir John Franklin's intention, if foiled at one point, to try in succession all the probable openings into a more navigable part of the Polar Sea. The range of coast is considerable in which memorials of the ships' progress would have to be sought for, extending from Melville Island, in the west, to the Great Sound, at the head of Baffin Bay, in the east." The same authority told Lady Franklin, that Sir John Franklin mentioned to him that, if he were baffled in everything else, he might perhaps look into the Sounds north of Baffin Bay before he returned home.

The intense anxiety and apprehension now so generally entertained for the safety of Sir John Franklin, and the crews of the *Erebus* and *Terror* under his command, who, if still in existence, are now passing through the severe ordeal of a fourth winter in those inclement regions, imperatively calls for every available effort to be made for their rescue from a position so perilous; and as long as one possible avenue to that position remains unsearched, the country will not feel satisfied that everything has been done which perseverance and experience can accomplish to dispel the mystery which at present surrounds their fate.

The plan I propose is neither difficult nor expensive in the accomplishment. Jones Sound is within the short distance of about 100 miles of Lancaster Sound; and Smith Sound is scarcely as much farther north of Jones Sound. The *North Star* ought to arrive there about the beginning of August, which month, with part of September, would leave nearly two of the best months of the year for the examination of one or both these Sounds to their probable termination in the Polar Sea. Jones Sound, with the Wellington Channel, on the west, may be found to form an island of the land called "North Devon." All prominent positions on both sides of these Sounds should be searched for flag-staves and piles of stones, under which copper cylinders or bottles may have been deposited, containing accounts of the proceedings of the missing expedition; and if successful in getting upon its track, a clue would be obtained to the fate of our gallant countrymen.

The s.arching party should commence its return in time to

reach the entrance of Jones Sound at an appointed time and place, at which the *North Star* should be directed to call, after she had delivered her stores for the ships in Lancaster Sound. The latter part of September would be early enough for her final departure (so as to secure her from being beset for the winter), as that month is well known to be the best period of the year for navigating Baffin Bay.

Having already twice volunteered my services to the Admiralty as long ago as the year 1847, to be employed in the general search for the lost Expedition, I need scarcely add how happy I shall be to conduct such a Boat Expedition as the one I have proposed. All that I should require for the performance of such a service would be an open boat—a whale boat would, perhaps, be the best, with a tent and stove, and the requisite equipment for her crew, six in number.

And further, should it be found necessary to continue the search in the following year, I am ready to winter on the coast in a log-hut, supplied with sufficient fuel, provisions, and clothing for the existence of my small party through a polar winter, or on board the *North Star* should it ultimately be deemed desirable that vessel should remain out.

Should this outline of my plan meet with approval, I am prepared to furnish a more detailed statement.

(Signed) R. McCormick,
Surgeon, R.N.

11, Craven Street, 21st April, 1849.

No. 3.

Walpole Lynn, Norfolk,
18th April, 1849.

Dear McCormick,—I need scarcely say that I greatly admire the zeal which has prompted you to propose an additional plan for obtaining information respecting Sir John Franklin's Expedition, and that I consider any proposition coming from one so well acquainted with the polar regions as yourself well worthy of attention, more especially when you offer your own services in putting it into execution.

I will now give you my deliberate opinion as to the utility and practicability of the plan you have submitted to me.

There can be no doubt as to the importance, and considering our increasing anxiety respecting Sir John Franklin, I could almost say the necessity, of examining the various sounds and inlets between Lancaster Sound and the head of Baffin Bay.

More especially as Jones Sound is said to have been recently entered by a whaler, and found to extend a considerable distance to the westward.

Among the probabilities to which we are now driven, there is none more likely than that Sir John Franklin may have tried some one of those inlets, after failing in Lancaster Sound.

Then, as to your plan of effecting this examination, I am decidedly of opinion that, so long as the summer remains open, by which I mean until young ice begins to form a serious impediment in shore, there is more to be done by a well-equipped boat than in any other way; both as regards certainty of progress and actual safety, more is to be done in close examination in a boat than in a ship, as I have more than once found by experience.

Two questions then arise—

1st. What time would you have for effecting this object, supposing you went out in the *North Star?*

2nd. What prospect of securing your retreat, or of wintering in safety?

I am decidedly of opinion that the *North Star* cannot for one moment be permitted to go out of her course (i.e. to be diverted from her main object of delivering provisions to the *Investigator* in order to land any resources for you at the mouth of Lancaster Sound, much less at any place to the northward of it).

I am more and more confident that the *North Star* will have little or no time left after delivering her stores (which is no easy job), and if she has, we have proposed that she shall be employed under Captain Bird to carry on the search; so that either in the case of her coming home this year or not, you could get no help from her; none, I mean, independently of her delivering of stores to the *Investigator*, or in such other place as circumstances may render necessary as the general depôt for Sir James Ross's Expedition.

It is therefore perfectly clear to my mind, that your prospect of being provided for during the winter must depend upon your reaching the *Investigator* or some known depôt of provisions either at Port Leopold, or some point on the south shore of Lancaster Sound, before the winter sets in. I feel confident that if you trust to any other resource, you will be disappointed.

It follows, then, in my view, that you must leave the *North Star* at the entrance of Lancaster Sound with what your boat will stow, and that you must return in time to some known depôt of provisions, independently of anything to be specially deposited for you, for it is impossible for the *North Star* to execute a third object this season.

If, therefore, the *North Star* should succeed in reaching Lancaster Sound (as I once did) in the early part of August, you might do a great deal in the six weeks following, and secure your own retreat; if much later, you could do so much the less, and perhaps incur so much the greater risk.

These are my general views, and I do not entertain them lightly.

Upon the whole, I do think it would be worth while to let you have a boat to make the attempt; this would cost little, in any way, even if it failed, and I should be glad to see so much ardour as you possess employed in this humane and noble cause.

I have, &c.

(Signed) W. E. PARRY.

R. McCormick, Esq.

No. 4.

MEMORANDUM enclosed in Mr. McCORMICK'S Letter of 1st January, 1850.

IN the month of April last, I laid before my Lords Commissioners of the Admiralty a plan of search for the missing Expedition under the command of Captain Sir John Franklin, by means of a Boat Expedition up Jones and Smith Sounds, volunteering myself to conduct it.

In that plan I stated the reasons which had induced me to direct my attention more especially to the openings at the head of Baffin Bay, which at the time were not included within the general scheme of search.

Wellington Channel, however, of all the probable openings into the Polar Sea, possesses the highest degree of interest, and the exploration of which is of paramount importance, I should most unquestionably have comprised within my plan of search, had not her Majesty's ships *Enterprise* and *Investigator* been employed at the time in Barrow Straits for the express purpose of examining this inlet and Cape Walker, two of the most essential points of search in the whole track of the *Erebus* and *Terror* to the westward; being those points at the very threshold of his enterprise, from which Sir John Franklin would take his departure from the known to the unknown, whether he shaped a south-westerly course from the latter, or attempted the passage in a higher latitude from the former point.

The return of the Sea Expedition from Port Leopold, and the overland one from the Mackenzie River, both alike unsuccessful in their search, leaves the fate of the gallant Franklin and his companions as problematical as ever; in fact, the case stands precisely as it did two years ago; the work is yet to be begun; everything remains to be accomplished.

In renewal of the search in the ensuing spring, more would

be accomplished in boats than in any other way, not only by Behring Straits, but from the eastward. For the difficulties attendant on icy navigation, which form so insuperable a barrier to the progress of ships, would be readily surmounted by boats; by means of which the coast-line may be closely examined for cairns of stones, under which Sir John Franklin would most indubitably deposit memorials of his progress in all prominent positions, as opportunities might offer.

The discovery of one of these mementos would, in all probability, afford a clue that might lead to the rescue of our enterprising countrymen, ere another and sixth winter close in upon them, should they be still in existence; and the time has not yet arrived for abandoning hope.

In renewing once more the offer of my services, which I do most cheerfully, I see no reason for changing the opinions I entertained last spring; subsequent events have only tended to confirm them. I then believed, and I do so still, after a long and mature consideration of the subject, that Sir John Franklin's ships have been arrested in a high latitude, and beset in the heavy polar ice northward of the Parry Islands, and that their probable course thither has been through the Wellington Channel, or one of the Sounds at the northern extremity of Baffin Bay.

This appears to me to be the only view of the case that can in any way account for the entire absence of all tidings of them throughout so protracted a period of time (unless all have perished by some sudden and overwhelming catastrophe).

Isolated as their position would be under such circumstances, any attempt to reach the continent of America at such a distance would be hopeless in the extreme; and the mere chance of any party from the ships reaching the top of Baffin Bay at the very moment of a whaler's brief and uncertain visit would be attended with by far too great a risk to justify the attempt, for failure would ensure inevitable destruction to the whole party; therefore their only alternative would be to keep together in their ships, should no disaster have happened to them, and by husbanding their remaining resources, eke them out with whatever wild animals may come within their reach.

Had Sir John Franklin been able to shape a south-westerly course from Cape Walker, as directed by his instructions, the probability is, some intelligence of him would have reached this country ere this (nearly five years having already elapsed since his departure from it). Parties would have been sent out from his ships, either in the direction of the coast of America or Barrow Straits, whichever happened to be the most accessible. Esquimaux would have been fallen in with, and tidings of the long absent Expedition have been obtained.

Failing in penetrating beyond Cape Walker, Sir John Franklin would have left some notice of his future intentions on that spot, or the nearest accessible one to it; and should he then retrace his course for the Wellington Channel, the most probable conjecture, he would not pass up that inlet without depositing a further account of his proceedings, either on the western or eastern point of the entrance to it.

Therefore, should my proposal meet with their Lordships' approbation, I would most respectfully submit that the party I have volunteered to conduct should be landed at the entrance to the Wellington Channel, or the nearest point attainable by any ship that their Lordships may deem fit to employ in a future search, consistently with any other services that ship may have to perform; and should a landing be effected on the eastern side, I would propose commencing the search from Cape Riley or Beechey Island in a northerly direction, carefully examining every remarkable headland and indentation of the western coast of North Devon for memorials of the missing Expedition; I would then cross over the Wellington Channel, and continue the search along the northern shore of Cornwallis Island, extending the exploration to the westward as far as the remaining portion of the season would permit, so as to secure the retreat of the party before the winter set in, returning either by the eastern or western side of Cornwallis Island, as circumstances might indicate to be the most desirable at the time, after ascertaining the general extent and trending of the shores of that island.

As, however, it would be highly desirable that Jones Sound should not be omitted in the search, more especially as a whaler, last season, reached its entrance and reported it open, I would further propose, that the ship conveying the exploring party out should look into this opening on her way to Lancaster Sound, if circumstances permitted of her doing so early in the season; and, if found to be free from ice, the attempt might be made by the Boat Expedition to push through it to the westward in this latitude; and should it prove to be an opening into the Polar Sea, of which I think there can be little doubt, a great saving of time and distance would be accomplished. Failing in this, the ship should be secured in some central position in the vicinity of the Wellington Channel, as a *point d'appui* to fall back upon in the search from that quarter.

(Signed) R. McCormick, R.N.

Twickenham, 1st January, 1850.

No. 5.

COPY of a LETTER from Mr. McCORMICK to the Secretary of the Admiralty.

11, Apsley Cottages, Twickenham Green,
20th February, 1850.

SIR,—I beg leave to transmit herewith, for the approval of my Lords Commissioners of the Admiralty, a list of the crew, gear, and provisions and clothing requisite for the equipment of the Boat Expedition, which I have volunteered to conduct in search of her Majesty's ships *Erebus* and *Terror*, under the command of Captain Sir John Franklin.

The boat I should prefer for this service would be one similar, in the materials of its construction, to the boat used by Sir Edward Parry, in his attempt to reach the North Pole in the year 1827; but this I must leave to the superior judgment of their Lordships. The timbers in that boat were of tough ash and hickory, with Mackintosh's waterproof canvas, and oak and fir planking over all, and having a runner on each side of the keel shod with smooth steel. The boat I would propose should be built after the model of a whale-boat, twenty-five feet in length and five feet beam in the extreme breadth; the crew to consist of a petty officer, a carpenter, and four able seamen.

The route which I am the most desirous and anxious to follow is by the Wellington Channel; so strongly impressed am I with the conviction that it affords one of the best chances of crossing the track of the missing Expedition, for the reasons already stated in my plan, now under their Lordships' consideration.

To carry out this plan efficiently, the boat should be dropped by the ship conveying the searching party out at the entrance to the Wellington Channel in Barrow Straits; from this point one or both sides of that channel and the northern shores of the Parry Islands might be explored as far west as the season would permit of. But should the ship be enabled to look into Jones Sound, on her way to Lancaster Sound, and find that opening free from ice, an attempt might be made by the Boat Expedition to push through it into the Wellington Channel. In the event, however, of its proving to be merely an inlet, which a short delay would be sufficient to decide, the ship might perhaps be in readiness to pick up the boat on its return, for conveyance to its ultimate destination through Lancaster Sound, or as a precaution against any unforeseen separation from the ship, a depôt of provisions should be left at the entrance to Jones Sound for the boat to complete its supplies from, after accomplishing the exploration of this inlet, and to afford the means, if compelled from an advanced period of the season or other adverse circumstances, of reaching some place of refuge, either on board a whaler or some one of the depôts of provisions on the southern shores of Barrow Straits.

I have, &c.

(Signed) R. McCORMICK, R.N.

Equipment for the Boat.

Boat's gear, awning, tarpaulins, &c.
Carpenter's tools and vulcanized Indian rubber for repairs.
Six fowling-pieces and ammunition.
Arm-chest and magazine.
Harpoons and net.
Compass, chronometer, and quadrant.
Thermometers and aneroid barometer.
Box of stationery.
Tent and Smith's Orion belts.
Halkett's boats (large and small sizes).
Two of Sir Edward Parry's sledges.
Cooking apparatus, and knives, forks, and spoons.
Gutta pe cha cups and plates.
Weighing dial and measures.

Provisions.

(Rations for each man per diem.)

Pemmican	12 oz.
Preserved meats, soups, and vegetables	12 "
Biscuit	1 lb.
Cocoa powder, sweetened	1 oz.
Chocolate	1 "
Tea	0½ "
Sugar	1 "
Rum	0½ gill.
Tobacco	0½ oz.
Spirits of wine for fuel	1 pint.

Clothing.

Fur caps and south-westers, of each	7 no.
Fur dresses for sleeping in	7 "
Suits of pilot cloth	7 "
Cloth boots and moccasins	14 pairs.
Mittens and stout stockings	14 "
Guernsey frocks and flannel shirts	14 "
Flannel drawers and comforters	14 "
Blanket bags for sleeping in	7 "

(Signed) R. McC.

No. 6.

MEMORANDUM by the Hydrographer of the Admiralty on Mr. McCORMICK'S proposed Boat Expedition.

DR. McCORMICK has shown so much heart and perseverance in urging his project for the relief of the *Erebus* and *Terror*, that there can be no doubt that he would execute it with commensurate zeal and resolution; and though it does not appear to me that Jones Sound or the Wellington Channel are the most likely places to find those ships, yet in the fifth year of their absence every place should be searched, and I therefore submit that this plan would cost but little, as a rider upon some other Expedition by the eastern route. If their Lordships should consent to Captain Penny's offer of proceeding to Lancaster Sound in his whaling vessel, perhaps the doctor might be despatched with him, according to the position they might find occupied by the ice: they would be able to determine at which point of his proposed circuit it would be most prudent for him to land, and they would arrange at what place he should be picked up.

(Signed) F. B.

No. 7.

4, Elm Court, Temple,
31st March, 1850.

DEAR SIR FRANCIS,—If Mr. McCormick's offer of service could be considered as an application for employment, I should have declined from the first to have meddled with it, feeling, as I do, that upon the present administration I have no claim whatever, and that so long as I remain an independent member of Parliament I am not likely to find myself otherwise situated in respect of any administration.

Humanity and bravery are qualities which all men of principle must appreciate; and when, as in McCormick's case, they are tempered by prudence and guided by long practice, all men ought, in my opinion, to unite in cordial efforts to promote the success of what they undertake.

You do not need testimony in this instance of humanity and courage; your own heart will bear him witness. Of the other essential qualities for success in the scheme of search which he proposes, the valuable reports of experienced officers, certifying their approbation of his plan, and, with that plan, laid before Parliament, will more than suffice to convince a mind constituted like your own. Nevertheless, I may recall to your recollection the *private* opinion of Sir James Ross himself—certainly no *patron* of McCormick—and the opinion of Commander Phillips, his companion in the boat expedition of survey on the coast of Kerguelens in 1840, the purport of which I quoted to you in a recent conversation. Captain Austin, too, *although the plan had never been referred to him*, on Thursday last told McCormick in Captain Becher's presence, that he was most favourably disposed towards it, and wished it success. I mention this because I know that ill-informed persons are under a different impression as to Captain Austin's inclination to forward plans of search independent of that plan which is immediately under his control and management.

The Board over which you preside have reconsidered the matter, and permitted me to inform McCormick that his plan *shall* be tried. Admiral Dundas went so far as to assure me, at a subsequent interview, that the *Intrepid* was to take him and his boat. But the boat CANNOT be built in LESS than a *fortnight*, and it MAY take a MONTH, according to the state of the dockyard. Yet the order has not yet been given, nor has McCormick ever received any notification of his appointment, nor, indeed, any communication whatever from the Admiralty. In the mean time, the season advances, the *Intrepid* is getting every day more ready for sea, and the difficulty begins to be apprehended that, when the order is at length given, there will not be time to build the boat; or if built, that there will not be room on board the *Intrepid* to receive it. This boat is, however, the root of the matter, the plan being without it impracticable and worthless. Indeed, I do not see that it would be very wise for McCormick to go out at all without his boat. He can do nothing if he goes, and, as it strikes me—remembering that Sir William Burnett's influence (or some other) has hitherto deprived him of the step to which his Antarctic services in 1839-43 on board of this very *Erebus* entitled him so well—it will not do *for him to go out* with the tolerable certainty of coming back *unsuccessful*. Reputation to any officer—to him more especially—must be all in all, where promotion is in question. Pardon the freedom of these remarks, which are not intended, I beg you to believe, as arguments for McCormick, but, rather, for poor Franklin. I am firmly persuaded that if the plan be *fairly* tried, it will save all that is to be saved of the lost expedition, I therefore urge it upon your goodness and zeal in this uncourteous fashion. I have a peculiar motive to stimulate the anxiety which I share with others. Sir John Franklin was for a few months friendly towards me, and then he and I had differences of a political nature, and then we quarrelled, and I thought that he wronged me. Do not let me fail in my endeavour to serve him at his need. I am sure that he, if he returns, will appreciate the service.

Believe me, dear Sir Francis,
Yours truly,
(Signed) T. CHISHOLM ANSTEY, M.P.

Rt. Hon. Sir Francis Thornhill Baring, Bart., M.P., &c.

No. 8.

4, Elm Court,
2nd April, 1850.

DEAR SIR FRANCIS,—I have to thank you for your early reply to mine of the 30th (misdated the 31st) ult.

Permit me to add to the remarks contained in my former letter that this is a service in which unfortunately no account can be taken of the risks which the volunteers incur. To make an effectual search there must be dreadful risks ventured. The question should be, is such or such a plan of search likely to be effectual? The more likely, the greater risk, perhaps.

Tried by this test, McCormick's plan appears a proper one for your adoption, and of the risk he and his comrades alone are to judge. Let me, however, say that perhaps you overrate the amount. There must be a receiving-ship, a *point d'appui*, somewhere stationed for the support and relief of the smaller vessels. Why may not his boat obtain the benefit of her being there? or, why may he not, at the very worst, reach Port Leopold, where there are twelve months' provisions, and the steamer left by Sir James C. Ross last year? I know, that he accounts the risk as nothing in comparison of the hope of success.

The disinclination of the Board to try his plans was removed by subsequent communications, unofficial, doubtless. Admiral Dundas' promise to give him a berth on board of the *Intrepid*, and a boat, was subsequent to the letter of the Board, declining that offer of his services.

I have thrown these hasty observations together in the purpose of enabling you better to judge the present state of the case when you come to consult your colleagues,

And I am, dear Sir Francis,

Very faithfully yours,

(Signed) T. CHISHOLM ANSTEY, M.P.

Rt. Hon. Sir Francis T. Baring, Bart., M.P., &c. &c.

No. 9.

Mr. MCCORMICK, to the Secretary to the Admiralty.

Apsley Cottages, Twickenham Green,
27th November, 1851.

SIR,—May I request that you will be pleased to acquaint my Lords Commissioners of the Admiralty that I am ready and willing as ever to conduct a "boat and sledge expedition" in search of her Majesty's ships *Erebus* and *Terror*, under the command of Captain Sir John Franklin.

Having been the *first* to propose the mode of search by "*boat and sledge*," as well as the *first* to point out the *Wellington Channel* as the course taken by the missing expedition in the attempt to accomplish the "North-west passage," now placed almost beyond a doubt by the traces found at Cape Riley and Beechey Island—the very spots named in the plan I had the honour of submitting to their Lordships on the 1st of January,

1850, as the first to be searched for memorials, and the most likely places for striking upon the track of the missing ships—affords the surest guarantee for the successful execution of a project so auspiciously planned.

I have, &c.,
R. McCormick,
Surgeon, R.N.

No. 10.

Plan of Search, by Boat and Sledge, for the Rescue of Captain Sir John Franklin and the Crews of Her Majesty's Ships *Erebus* and *Terror*, or the discovery of their Fate.

Expedition after expedition, both by sea and land, have been sent forth by England and by America, by public and by private enterprise, in search of our lost countrymen, and returned again and again, leaving their fate as inexplicable a mystery as ever.

Yet, strange enough, not one of those expeditions have explored *Smith Sound*, at the head of Baffin Bay, looked into it, or even made the attempt. Although, next to Wellington Channel, the most promising and important opening to the Polar Ocean, within the icebound recesses of which there can now scarcely be entertained a rational doubt that the ill-fated *Erebus* and *Terror* have been inextricably beset, or wrecked amongst the heavy packs and archipelago of islands by which, in all probability, that ocean is encumbered.

My own opinion has ever been that Sir John Franklin went up the Wellington Channel, and, consequently, the surest way to find him would be, not only to follow upon his track up that channel, and to the northward and westward of the Parry group of land, but also to meet him in any retrograde movement he might be compelled to make to the eastward, should his ships, in the attempt to get to the westward, be driven by the strong currents from the *north-west* to the meridian of the Sounds at the northern extremity of Baffin Bay; a by no means improbable event, and one that should not be lost sight of. If Smith and Jones Sounds should prove to be openings into the Polar Ocean, as I long ago anticipated, they would offer the readiest means of exit to either ships or boats, with the prospect before them of falling in with some whaler.

It was under this impression that I offered my services, as long ago as the year 1849, to go out in her Majesty's ship *North Star*, to conduct a boat expedition up those Sounds, volunteering at the same time to winter on the coast in a log-hut, if provided with a whale boat and half a dozen hands, &c. This offer, although most favourably reported upon by Captain Sir Edward Parry, was declined by the Lords Commissioners of the Admiralty.

On the 1st of January, 1850, I laid another plan before their Lordships for the exploration of the Wellington Channel by boat and sledge. In that plan I made the following remarks:—"Wellington Channel, however, of all the probable openings into the Polar Sea, possesses the highest degree of interest, and the exploration of which is of paramount importance," &c. "That Sir John Franklin's ships have been arrested in a high latitude, and beset in the heavy Polar ice northward of the Parry Islands, and that their probable course thither has been through the Wellington Channel," &c.

"I would propose commencing the search from *Cape Riley* or *Beechey Island*, in a northerly direction, carefully examining every remarkable headland and indentation of the western coast of North Devon for memorials of the missing expedition," &c.

Since these remarks were written, the cape and island have become well known as the first winter quarters of the unfortunate ships, and it may appear a striking coincidence that I should have named these very spots. The motive for my doing so, however, is easily explained. The very suitable position of the little bay at Cape Riley, so well protected by Beechey Island, which formed a natural fender for keeping off the heavy floes and pack ice, at once pointed it out as a most desirable harbour for the ships at the very threshold of their enterprise, and could not fail, on glancing over the chart, to rivet the attention of an experienced and practised eye.

Again, in a subsequent letter to the Board, dated 20th February, 1850, I stated, "The route which I am the most desirous and anxious to follow is by the Wellington Channel, so strongly impressed am I with the conviction that it affords one of the best chances of crossing the track of the missing expedition, for the reasons already stated in my plan now under their Lordships' consideration."

In these views of the position I had assigned to the missing expedition I believe I at the time stood alone. The generally received opinion having been that Sir John Franklin's ships had been arrested in the ice to the southward and westward of Cape Walker or Melville Island. The results of the late searching parties have, however, proved beyond a doubt the correctness of my own views, even to the finding of traces, if not a memorial (which, however, I believe, yet remains to be discovered), at Cape Riley and Beechey Island.

Although none could have felt more keenly than I did the disappointment in not having been permitted to carry out either of my projects, feeling as I then did so confident that success would have crowned my efforts, I nevertheless, even now, am as sanguine as ever that it is not yet too late to save some gallant fellows, if not all, from a lingering fate, too fearful to dwell upon —from a living tomb.

My own personal knowledge of the resources available for sustaining life within the Arctic regions forbids the thought that a hundred and thirty fellow-beings in the full vigour of manhood have already succumbed under the effects of cold, famine, or disease, without one individual being left to tell the sad and melancholy tale.

The "snow hut" would afford them shelter from the weather, the skin of the seal protection from the cold, its blubber light and fuel. The "Andromeda tetragona," a plant of the heath tribe, wi lely spread over arctic lands, and which I have myself gathered in the northernmost known land, Spitzbergen, where it grows in considerable abundance, offers another source from which fuel may be obtained. Vast flocks of waterfowl which annually migrate to their breeding-places in the very depths of the arctic solitudes, where they can rear their young unmolested by man or beast, probably beyond even the range of that restless wanderer of the snowy wastes, the arctic fox—these birds would be easily captured whilst moulting and unable to fly, and with their eggs furnish a wholesome supply of food for each succeeding winter's store. Scurvy is the foe, after all, the most to be dreaded, and progressively so with the lapse of time, and gradual decline of the vital powers, but even this scourge, whilst it sweeps off the despondent and indolent, oft spares the buoyant and energetic.

My firm belief that the crews of my old ships the *Erebus* and *Terror*, or a remnant of them, are still in existence, is founded on some years' personal experience in frozen climes, both Arctic and Antarctic, and my observations as a naturalist on the habits and instincts of animals, with their geographical distribution over the surface of the globe, from pole to pole, leads me to the conclusion that the means of sustenance will not be wanting. Under this conviction, my enthusiasm in this noble cause will never cease to prompt me to come forward to their rescue on every occasion that may offer for carrying out my plan of relief, till the problem has been solved that shall decide their fate, and not till then.

Once more, therefore, and for the fourth time, I may be allowed to call the attention of my Lords Commissioners of the Admiralty to a reconsideration of my plan, and if above four years of unceasing and unwearied application to be employed in the Franklin search be any proof of zeal, perseverance, and devotedness of purpose, and these qualities, when backed by experience, considered fitting qualifications for such an undertaking, I trust that their Lordships will permit me to have some share in the search on its renewal in the forthcoming spring.

All I ask for is a whale boat and sledge, manned by six hands, with the requisite equipment of stores, fuel, provisions, clothing, &c., and the command of the party, with which it was my original intention to have proposed proceeding direct to Smith Sound, and devoting the ensuing autumn to the exploration of that inlet as far up as the season would admit of, wintering there in a log-hut taken out for the purpose, so as to be enabled in the following spring to extend the search over the ice by sledging; and in the event of Smith Sound opening into the Polar Ocean, which I believe it does, and the heavy swell setting out of it (as indicated in the Admiralty chart of Baffin Bay, published in the same Return to the House of Commons in which my own former plans appeared, and dated March, 1850), is strongly in support of such an opinion, and by shaping a westerly course, a junction might possibly be effected with the searching parties employed up the Wellington Channel.

I find, however, that the Arctic Council have recommended that the future search shall be exclusively confined to the Wellington Channel, and that a squadron of ships be sent out in that direction. Moreover, an objection might be made to the attempt to explore Smith Sound from Baffin Bay, on the ground that the entrance to it has never been seen clear of ice, but were such the fact I know not how we are to account for the heavy swell.

Under these circumstances I most willingly volunteer my services to go out in any one of the vessels to Wellington Channel, there to commence the search in the reverse order, round *Cape Sir John Franklin*, northward or eastward, as the land may trend, exploring in the direction of the meridian of Smith and Jones Sounds for any corresponding openings to the Polar Ocean, into which the missing ships may have been driven under the influence of adverse winds and currents, whilst helplessly beset in heavily and closely packed ice.

Such a branch enterprise, carried out at the same time with the still more important one to the westward round Cape Lady Franklin, and which will doubtless be the main object of the next general expedition, would by providing for every contingency, promise the best possible chance of restoring to their friends and country all that remains to be saved of our brave and enterprising countrymen.

R. McCormick,
Surgeon, R.N.

Twickenham, 20th January, 1852.

No. 11.

LAST PLAN OF SEARCH FOR THE DISCOVERY OF THE FATE OF SIR JOHN FRANKLIN'S EXPEDITION, laid before the Lords Commissioners of the Admiralty and the Royal Geographical Society, by ROBERT MCCORMICK, F.R.C.S., R.N., January 6th, 1857.

Reasons for a renewal of the search for further traces of the Franklin Expedition, and the various routes by which that search may be carried out, considered in reference to the most eligible plan for reaching the area to be explored, comprising King William's Land and the coasts adjacent.

THE accidental discovery of the relics left in the hands of the Esquimaux, at the mouth of the Great Fish River, by that gallant band of toilworn and famishing men, the forlorn hope of 40 out of the 138 noble fellows that constituted the crews of the ill-fated *Erebus* and *Terror*, gives a new and important feature to the search,—a definite spot on which to concentrate our efforts; a clue which, if zealously and judiciously followed up, cannot fail to lead to the discovery of the fate which befell our long-lost countrymen.

What a startling narrative of adventures and discovery, what a thrilling tale of suffering and woe, may not the recovery of a record of their wanderings unfold; though the hand that penned it, and the hands that buried it, be long ago cold and inanimate as the cairn of stones beneath which that record may lie concealed amid the frozen, snow-clad cliffs overhanging the ice-bound shores of King William's Land, or the banks of the Great Fish River? The probability of finding such a document would, in itself, justify the prosecution of a further search, were there no other reasons for continuing it. In the present crisis, however, it cannot be said to be the only one. The relatives and friends of the lost ones, still poised between hope and despair, in that painful position of suspense and anxiety, worse than the dread reality itself, cling, naturally enough, to what I fear must now be considered delusive hope that survivors will be found—a hope, nevertheless, held out not by the inexperienced and unreflecting alone, but by those whose position and judgment give influence and weight to their opinions, and whose sanguine temperaments make them reluctant to resign that hope so long as mystery remains, albeit the twelfth year of this sad mystery is drawing to a close.

In again pressing for another search, I need scarcely say that I do so solely from a sincere and firm conviction in my own mind of the necessity for it, ere this much vexed question can possibly be set at rest for ever. As a Surgeon in the Royal Navy, I presume no selfish or interested motives can be imputed to me, seeing that I belong to a class hitherto excluded from sharing in the promotions which have been so liberally bestowed on officers of every other grade, on the return of each successive expedition: albeit the Surgeon, no matter whether he conducted a successful expedition himself, or was one of the discoverers of a Polar Passage, a Surgeon he remains.

The area to be searched for records is now reduced to within very circumscribed limits, and may be comprised within the parallels of 67° and 74° of latitude, and the meridians of 94° and 100° of longitude. Within this space lie Peel Sound, King William's Land, the western coast of Boothia, and the estuary of the Great Fish River, King William's Land itself occupying about 4 degrees of longitude, and 1½ degrees of latitude. This area of search may be approached from each of the four cardinal points of the compass.

From the west, by Behring Strait; from the east, by Hudson Bay; from the north, by Barrow Strait; and from the south, by the Great Fish River.

The first route, by Behring Strait, may be at once discarded, as it now *too late* in the season for any attempt in that direction. The vast distance also to the sphere of action is, of itself, a great objection to this route; for, under the most favourable circumstances, not less than nine or ten months must be sacrificed on the passage out; and although a ship's reaching even King William's Land in the same season comes within the range of possibility, it needs no very profound knowledge or experience in ice-navigation to foresee that this must greatly depend upon a favourable season. Nothing in nature is more capricious than the movements of ice, under the influence, at all times, of winds and currents. Strange fatality! that the *Enterprise's* sledge parties should have been within the short distance of *forty-five* miles of King William's Land without reaching its shores,—shores now so replete with interest. Yet in the vicinity of Cambridge Bay, their winter quarters, *only* 120 miles off, a door-hatch and a piece of iron, supposed to have belonged to the *Erebus* and *Terror*, appear to have been found.

The second route, by Hudson's Bay, presents three openings; two of them, Chesterfield Inlet and Wager River, run up in the direction of the Great Fish River, but the intervening space, though of no great extent, which separates them from the latter river, is said to be rough ground, unfavourable for travelling. The third, Repulse Bay, involves a long overland journey for the transit of a boat of a size required for the navigation of the western sea, first, across the isthmus into Committee Bay, and from thence over Franklin Isthmus to Castor and Pollux River.

The third route, by Barrow Strait, offers the choice of three avenues to King William's Land and the Great Fish River—by

Melville Sound, Peel Sound, and Regent Inlet. Melville Sound, situate to the westward of Cape Walker, in the longitude of 105° west, is, from the report of the sledging party that explored it, *not navigable* for ships at any period of the year, being beset by tremendous ice, frozen to the bottom, and rising in hummocks from fifteen to twenty-five feet in height, the land low, with off-lying shoals at the bottom of the Sound, about fifty miles of which was not explored, but supposed to be *continuous*. Peel Sound, lying between North Somerset and Cape Walker, in the longitude of 96°, was also examined by sledging parties, down to the latitude of 73° (180 miles beyond this remain unexplored), holds out little better prospect for the passage of a ship, from the character given in the report—that it is filled with undisturbed, smooth floe-ice, aground, exhibiting no appearance of pressure or tide-marks along its flat shores, and rarely, if ever, open to navigation. The smooth ice, however, would be favourable for sledging over, with a ship secured in one of the inlets of the coast of North Somerset, for the party to fall back upon.

Regent Inlet, in the longitude of 90° west, appears to me the most promising of all the avenues of approach to the area of search. We know that this inlet is *really* navigable for ships, Sir Edward Parry having, in the month of August, in the year 1819, passed as far down it, with the *Hecla* and *Griper*, as Cape Kater, about the latitude of Brentford Bay and Bellot Strait; and the *Victory*, in the autumn of the year 1829, reached Felix Harbour, at the bottom of the Gulf of Boothia.

I would, therefore, suggest that a small vessel, of some fifty tons burthen, fore-and-aft rigged, with a crew of a dozen Greenland seamen, and two mates, besides the officer in command, stored and provisioned for eighteen months, should leave this country about Midsummer Day, proceed down Prince Regent Inlet as far as Brentford Bay, and, on the vessel being secured there, a boat and sledge expedition pass at once through the Strait of Bellot into the Western Sea, explore the coast of Boothia down to the magnetic pole, from thence crossing over to Cape Felix, circumnavigate the shores of King William's Land by boat or sledge, as the presence of water or ice may indicate; and if successful in discovering a record or other traces of the lost expedition, and the season an open and favourable one, might *possibly* return to England the same year.

The vessel I have proposed may appear small for such an enterprise. Baffin, however, in the old *Discovery*, a vessel of not greater tonnage, made the circuit of the bay which bears his name, and accomplished more in a few months than has been done since by more ostentatious expeditions. The advantages of a small vessel over a large one are manifold;—her smaller draught of water, her less liability to injury in a nip, the facility with which she may be worked in narrow leads amongst ice, or beached to repair damages, a smaller crew to provide for, great saving of expense in the general equipment, and lastly, in the event of any disaster to the ship, the smaller the party, the better the chance of escape

Within the area to be searched for records, I do not for one moment expect to find either the *Erebus* or *Terror;* I should just as soon think of looking for them at the South Pole, as at King William's Land. I am, nevertheless, fully aware that this is by no means the general impression, and that I am making a bold assertion. I would ask, however, how these ships got there, if Peel and Melvil e Sounds are *never navigable* at any season of the year? for such is the report of the officers who explored these openings. Through what channel, then, could the ships have passed, unless they had *anticipated* the discovery of the North-west Passage by penetrating *Prince of Wales' Strait?* What were they doing during the long interval of *four* years, unaccounted for, from their departure from winter quarters, at Beechey Island, to the discovery of the relics at the mouth of the Great Fish River? Is it probable that in any position these ships could have been placed in, *south* of Barrow Straits,—the distance is not so great,—*four years* would have been allowed to elapse without an attempt to communicate with the Hudson's Bay settlements? and if such an attempt had been made, can we reasonably suppose that out of the two fine ships' companies, numbering 138 men, for years inured to hardship and exposure, not one should have succeeded in the undertaking?

It is more, I think, than ever probable that Sir John Franklin, after making every effort to carry out his Instructions, which he most assuredly would do, as long as there was a chance of doing so, finally, finding himself baffled in all his attempts to penetrate those impracticable openings to the southward and westward, attempted the *second course* pointed out to him by his Instructions,—the *Wellington Channel*—which, with Smith and Jones Sounds, I long ago anticipated would be found to open into that vast Polar Ocean, the existence of which the late Arctic explorations of the Americans have *confirmed*, which doubtless forms *the* Polar passage uniting the Atlantic and Pacific Oceans, not by narrow and tortuous straits, as in the more southern latitudes, but by an ocean uniting oceans—as the greatest of all Arctic geographers, the talented and immortal Sir John Barrow, the historian of Arctic discovery, long ago announced as a Polar basin.

In so vast a region, extending over some seventy degrees of longitude, from Wellington Channel to Behring Strait, we know not what discoveries may have been made in that four years of mystery. The Pole itself may have been reached, for aught we know to the contrary, ere the catastrophe happened to the ships, which drove the forlorn hope of forty men, as a last resource, to reach the shores of America, or perish by the way. I believe that they came done Wellington Channel, over the floe, which, on breaking up, would leave no trace behind to mark their weary way. That, revisiting their old winter quarters at Beechey Island, they found the boat which they were seen dragging along when the Esquimaux fell in with them at the mouth at the Great Fish River. That Sir John Franklin, before he launched his ships upon the dangerous navigation of an unknown and ice-encumbered sea, would, as a precautionary measure, leave a boat and cache of provisions behind him to meet any emergency arising out of disaster to his ships, seems to me more than probable. Sir Edward Parry, I know, did this, in his attempt to reach the North Pole in the year 1827, having myself assisted in hauling the boat over the ice, which was left, with the cache of provisions, on Red Beach, Spitzbergen, when the *Hecla* was beset in the pack-ice. When I first saw the tent-circles at Cape Riley, it struck me that the site selected was a very out-of-the-way one for making magnetic observations, with the ships lying at the very opposite extremity of the Bay, and having above three miles of rugged, hummocky ice intervening between them and Cape Riley, when a far more eligible spot presented itself on the narrow spit of shingle on which the graves are situated, dividing Erebus and Terror Bay from Union Bay, and close to the anchorage of the ships. Here, indeed, was found the ground plan of a building, which I think had obviously been the magnetic observatory.

These remarks are not hazarded without some little acquaintance with the subject to justify them, having myself been engaged in the voyage to the Antarctic seas, undertaken for magnetic purposes, when this same *Erebus* had been my ocean home for four long years. The "tent-circles," both at Cape Riley and at the south-eastern extremity of Beechey Island, appeared to me more likely to have formed the temporary encampment of a travelling party, watching for a favourable opportunity of crossing Barrow Strait. Prince Regent Inlet and Peel Sound, being about equi-distant, would, either of them, lead a party to the Great Fish River; and the cairn of empty preserved meat canisters found on Beechey Island may have been erected by that party to beguile the weary hours spent in waiting for a passage across the Strait.

In conclusion, I may add that some such a small sea expedition as the one I have proposed, down Regent Inlet and through the Strait of Bellot, in conjunction with an overland one down the Great Fish River, would appear to me to combine more of the elements of success than any single expedition, though of more ostentatious pretensions, whether by sea or land.

The only apology I have to offer for having entered so

elaborately upon the subject of this hapless search must rest upon the deep feelings of interest that have actuated me, for years past, in my endeavours to dispel the mystery which hangs over it. It is now ten years since I laid before the Board of Admiralty my first Plan of Search, down the Coppermine River, for the purpose of exploring the very region where relics of our unhappy countrymen have since been found. It is now eight years since I laid another Plan before the same Board, for exploring Jones and Smith Sounds into the Polar Ocean, since carried out by the Americans,—my own views *confirmed*, but the *honour won* for *another* country. Above three years have elapsed since my return from my boat expedition up the Wellington Channel, and two years, on the 23rd of last October, since I last volunteered my services to their Lordships, to follow up the search for relics, at the moment the discovery of them was first announced.

And now, in proposing my present plan of search by Regent Inlet and the Strait of Bellot, the greatest proof I can give of my devotion to this noble cause is the offer of my own services to conduct it, which I do most willingly, should that plan meet with the approval of my Lords Commissioners of the Admiralty, and long experience in icy seas, combined with an intimate acquaintance with natural history, geology, and geography; a profession superseding the necessity for the special appointment of a medical officer; and the having already had charge of an expedition in the search, be considered fitting qualifications for such an undertaking.

Pemba Cottage, Surbiton-on-Thames, January 6th, 1857.

R. McCORMICK, Surgeon, R.N.

REPLY.

Admiralty, 14th January, 1857.

SIR,—I have received and laid before my Lords Commissioners of the Admiralty your letter of the 6th inst., requesting employment with an Expedition in search of further traces of Sir John Franklin, and enclosing your reasons for a renewal of the search.

I am, Sir, your very humble Servant,

THOMAS PHINN.

R. McCormick, Esq., Surgeon, R.N.
Pemba Cottage, Surbiton, Surrey.

No. 12.

New York,
10th April, 1854.

MY DEAR SIR,—Your extremely kind note of the 22nd of March came to hand this morning, as also did a copy of your official report to the Admiralty of your explorations in Wellington Sound. Allow me to thank you sincerely for both. I have this evening read the report with deep interest, and must say that you give me a better idea of the mode of travelling in the arctic regions over ice, land, and water, than any I have seen. I like it in all respects, except the last paragraph. I must entirely dissent with you from your deductions and conclusions contained in that. In my opinion the evidence is stronger now than it was four or five years ago, that some of Franklin's party are still living. Take the facts which McClure gives us, he has been now three years in the Bay of Mercy, blocked up by the ice. Cannot he and his party live there as long as they could, or longer than they would, if they had been residing in London or New York, immersed in all the luxuries which those cities afford. McClure's evidence goes to show that abundance of food can be procured to sustain human life. I have no doubt by proper management he could obtain four times the quantity that would be requisite to sustain his party, and that, too, of a quality better adapted for an arctic climate than salt beef and pork. I think your remarks on the preservation of health in Polar climates will bear me out in this. I grant you if Franklin had to depend on the resources on board of his own ships, then is his safety "hoping against hope;" but I place him somewhere in the meridian of Behring Straits, and the parallel of 75°—similarly situated as McClure is, and where he can live as long as McClure can. When that part of the world is examined where Franklin is supposed to be, and he is not found, and no vestiges of ships, I then may begin to entertain the idea that he is no more on this earth. You must excuse me for differing in opinion, but that should not cause any unkind feelings, and I know it will not, for we both have the *cause* at heart.

My little expedition now out under the command of Dr. Kane, I hope will do something up Smith Sound. The latest date I have from him was the 23rd of July last at Upernavik; he was then only waiting for a wind to start; he would probably winter on the east side of that sound in as high latitude as he could obtain; and then with his dogs, sledges, &c., make land, ice, and water excursions. I think it quite possible he may go east instead of west, north of Greenland, Europe, and Asia, perhaps America also. Such a wish I expressed to him, and which he would follow if he found the current and ice setting to the eastward, which he probably would. You will please excuse me for making so long a communication; when I get on arctic subjects, I know not when to stop. Again thanking you for your note and report,

I am, with great regard,
Your friend,
(Signed) HENRY GRINNELL.

Dr. R. McCormick.

No. 13.

MY DEAR SIR,—I have read with very great interest the narrative you so kindly sent me of your expedition to Baring Bay. The whole subject has been full of attraction to me for many years, and I need not say that it is peculiarly so now.

I believe with you, and such, I apprehend, was my son's opinion, that Franklin passed to north-west from Wellington Channel, and that he must have been frozen in somewhere to the north of Melville Island. I have thought, however, deriving the idea, perhaps, from the circumstances of the long drift of the American Expedition, that he may have been afterwards carried eastward, upon some occasional breaking up of the ice-field under the influence of the same great current that Parry encountered on his ice-tramp, in which case, Dr. Kane may have the happiness of meeting the relics of the party. For I cannot think them dead, and were they my countrymen, I should hope to protest successfully against the Government's abandoning them.

My son's last letter to me, as he was leaving Upernavik on the 23rd of July, spoke of the probabilities of an open season, and the unusual quantities of ice in the Atlantic at the present time, confirms the idea. If so, I hope he has been able to press upwards through Smith Sound, well into the circumpolar Sea. The severity of last winter, however, augurs ill, I fear, of the chances of his early return, and it may be that we shall have to invoke the aid of Congress before another year to despatch a rescue-party on his track.

I thank you very sincerely for your politeness, and only regret that, my son's book being out of print, I cannot for a while offer you a copy of it in return.

I am, very respectfully,
Your faithful servant,
J. K. KANE.
Philadelphia, 14th April, 1854.

Dr. McCormick.

No. 14.

From Sir FRANCIS BEAUFORT.

St. Margaret's, October 12th, 1854.

DEAR DR. MCCORMICK,—I am really very much flattered by the copy you have sent me of your modest and quiet account of the little slice of Arctic Exploration which, after so much eager and zealous application, fell at length to your share.

Whatever reader follows your narrative, he will at least say—this man, from first to last, intended *well*, in every difficult case acted *well*, and told *well* both his reasonings and his doings. That you may reap the fruit of so much energy, and long enjoy the conscious approbation of so much generous devotion to humanity and friendship, are the earnest wishes of

Yours faithfully,
F. BEAUFORT.

No. 15.

From the late Hydrographer to the Admiralty, Rear-Admiral Sir FRANCIS BEAUFORT, K C.B.

11, Gloucester Place, August 15th, 1855.

DEAR DR. MCCORMICK,—After all that was done in the Arctic Seas by our ships, and after all that has since been done here by the Government and the public, I feel very sorry to learn by your note of yesterday, that no notice has been taken of your share of the Expedition, though sanctioned by the Board, and executed judiciously and successfully.

I am not sure in what manner their Lordships could most suitably mark their approbation of your singularly persevering zeal; but I am sure that where there's a will there's a way; and as we have never had a First Lord who is a quicker judge of real merit, or more disposed to inquire into that merit, so I feel satisfied that you should lay the whole matter before him, briefly and distinctly. Confinement to bed or sofa for six long months will not only explain my not being able to write conveniently with ink, but, also, my not being able to make any more active endeavours to serve you than by thus honestly expressing my sincere conviction that your efforts in the general cause were second to none in devotedness.

Yours very faithfully,
F. BEAUFORT.

No. 16.

From JOHN BARROW, Esq., F.R.S., to Sir JOHN LIDDELL, Director-General of the Medical Department of the Navy.

Admiralty, November 10th, 1855.

DEAR SIR JOHN,—You are well aware of the great interest I feel in my esteemed friend Dr. McCormick, and will, I am sure, pardon any endeavours on my part to befriend him. The only way I can hope to do so is by an appeal to the fountain-head of his profession. His services in the Antarctic Seas with Sir James Ross, at Spitzbergen with Sir Edward Parry, and latterly in a boat expedition in search of Franklin, are of a nature peculiar, as compared to those of any other officer in the Medical Department of the Navy, and of a very severe and arduous description.

It pains me to see him stand alone unrewarded for his toil, zeal, and spirited performances, and in very trying circumstances.

Interested as I have been in Arctic adventure, and conversant with the merits of all engaged in the Polar Seas, I do not hesitate to say, that none have acquitted themselves better; and convinced that it only requires a full knowledge of Dr. McCormick's case to insure some suitable reward being bestowed upon him. In making this appeal, I hope you will not think that I have presumed upon your friendship, as I do it from a sense of duty I owe to all concerned in these voyages, and none more than to my friend in whose behalf I now address you.

Believe me, &c.,

JOHN BARROW.

No. 17.

From the Vice-President of the Royal College of Surgeons JOHN F. SOUTH, Esq., to the Director-General, Sir JOHN LIDDELL.

Blackheath Park, May 21st, 1856.

MY DEAR SIR JOHN,—I understand my old pupil, Mr. Robert McCormick, is seeking your assistance towards attaining his promotion in the service with which he has been connected ever since he left St. Thomas's.

I feel much interest in his welfare, believing him to be an able and intelligent Surgeon, of great activity and industry, and quite capable of having much good work out of him yet.

If you can assist him, you will very much oblige me, and serve a very able and deserving officer.

Believe me,
My dear Sir John,
Yours very truly,
J. F. SOUTH.

No. 18.

From REAR-ADMIRAL HAMILTON, late Secretary to the Admiralty.

Maiden Newton, Dorset,
November 13th, 1858.

MY DEAR SIR,—I have written to Sir John Liddell, observing, as I am sure will be the case, that your services cannot fail to be well known to him.

I shall be glad to hear of your succeeding in your wishes. Your Antarctic service was a very responsible one; and in the Arctic searches there is abundant proof of your spirit of enterprise and professional skill.

Professor Owen's opinion and goodwill may be of service to you.

I remain,
Very faithfully yours,
W. A. B. HAMILTON,
Rear-Admiral.

R. McCormick, Esq.,
Surgeon, R.N.

No. 19.

CIRCULAR No. 349.

Admiralty, November 26th, 1858.

Qualifications for the rank of Medical Inspector of Hospitals and Fleets, and of Deputy-Inspector of Hospitals and Fleets.

With reference to Chapter III., Article 15, of Her Majesty's Regulations, which renders four years' service as Surgeon on board a commissioned ship necessary as a qualification for the rank of Medical Inspector of Hospitals and Fleets, or of Deputy Medical Inspector of Hospitals and Fleets, Her Majesty has been graciously pleased, by her Order in Council of the 13th inst., to direct that valuable service on shore, at home, or abroad, should also be allowed, in special cases, to count as time for rendering a Surgeon eligible for the before-mentioned ranks; and my Lords Commissioners of the Admiralty hereby make known the same for the information and guidance of all whom it may concern.

By command of their Lordships,
H. CORRY.

No. 20.

From JOSEPH HENRY GREEN, ESQ., President of the Royal College of Surgeons.

To the Right Honourable Sir John Pakington, Bart.,
First Lord of the Admiralty, &c., &c.

College of Surgeons, February 25th, 1859.

SIR,—I venture to solicit the permission of saying in behalf of Dr. McCormick, of the Royal Navy, that he is honourably known to the Royal College of Surgeons of England by numerous and valuable contributions to the Hunterian Museum, evincing great zeal and unusual ability in the pursuit of natural science.

In respect of his professional character, which is highly esteemed by all who have the honour of his acquaintance, I have no hesitation in recommending him respectfully, but earnestly, to the favourable notice of Her Majesty's Government.

I remain, Sir,
Your most obedient Servant,
JOSEPH HENRY GREEN,
President of the College of Surgeons.

No. 21.

Admiralty, June 13th, 1865.

SIR,—It being desirable to ascertain the state of efficiency for employment of the Medical Officers of the Royal Navy on half pay, I have to request that you will inform me whether you consider yourself fit for active service. An early reply is requested.

I am, Sir,
Your obedient Servant,
A. BRYSON,
Director-General.

R. McCormick, Esq.,
Deputy Inspector-General, R.N.,
22, Ridgway Place, Wimbledon.

No. 22.

Wimbledon, June 14th, 1865.

SIR,—In compliance with the request conveyed in your letter of yesterday's date, in reference to the state of efficiency for employment of the Medical Officers of the Royal Navy, I beg to acquaint you with my fitness and readiness to serve whenever called upon.

I am, Sir,
Your obedient humble Servant,
R. McCORMICK,
Deputy Inspector-General, R.N.

Dr. Bryson, C.B.,
Director-General of the Medical Department of the Navy.

No. 23.

Admiralty, July 6th, 1865.

SIR,—I have to request that you will appear at this office between eleven and three o'clock on Friday, the 14th instant, to be surveyed as to your fitness for active service.

I am, Sir,
Your obedient Servant,
A. BRYSON,
Director-General.

R. McCormick, Esq.,
Deputy Inspector-General, R.N.
22, Ridgway Place, Wimbledon.

No. 24.

Memorial of ROBERT McCORMICK, Deputy Inspector-General of Hospitals and Fleets, to His Grace the DUKE of SOMERSET, K.G., First Lord of the Admiralty, &c.

Junior United Service Club, July 18th, 1865.

Humbly showeth:

1. That your Memorialist in seniority has been forty-two years in the Navy, having entered the service and been promoted to the rank of Surgeon before any other officer on the active list of Medical Officers, whether Inspectors, Deputy Inspectors, or Surgeons; and was nine years Senior Surgeon to the Director-General himself.

2. That your Memorialist's services have been chiefly of a special nature, having been employed in three different expeditions of discovery to the Arctic and Antarctic regions. In these voyages, the duties of naturalist and geologist devolved upon your Memorialist, by whom they were performed, in addition to his duties as Chief Medical Officer, without emolument or

reward of any kind, thereby saving to the Government the expense which a specially-appointed naturalist and geologist must have incurred.

3. That your Memorialist served with Sir Edward Parry in his attempt to reach the North Pole in 1827; with Sir James Ross in his scientific expedition to the Antarctic seas, from 1839 to 1843; and lastly, in command of the Boat Expedition up Wellington Channel in search of Sir John Franklin in 1852-53, About seven years spent in icy seas, with long services in the Tropics, having served three times on the West India Station.

4. That your Memorialist has never refused active service, but has offered himself fit and ready to serve at all times; and on his return from the expedition in search of Sir John Franklin, was a volunteer for the Baltic and Black Sea Fleets at the commencement of the Crimean War.

5. That your Memorialist having so recently had the honour of attending one of your Grace's levées, and having met with so gracious and considerate a hearing, your Memorialist would not have so soon again pressed his claims, had not your Memorialist been unexpectedly summoned to the Director-General's Office on the 14th instant, to be surveyed as to his fitness for service at Hong-Kong, having already reported his fitness and readiness to serve in reply to a previous letter received from the Director-General, and repeated at the survey his readiness to go to Hong-Kong, or any other part of the world, by sea or land, to which he might be sent.

6. Your Memorialist, under these circumstances, feels confident that your Grace will not permit an old officer to be shelved without any other recognition of services, such as have been recorded at the Admiralty, than the tardy promotion to the grade of Deputy Inspector-General, nearly twenty years after the services rendered for it in the South Polar Expedition.

7. That no opportunity has been afforded your Memorialist of serving for routine promotion to the next rank, is no fault of his. And your Memorialist most respectfully and earnestly requests your Grace's attention to the Admiralty Circular of the 28th of November, 1858, which fully meets this difficulty in reference to the service time; by which a power is reserved to their Lordships in cases of distinguished and special services, to promote an officer *direct* from Deputy-Inspector to the Inspectorship.

8. Your Memorialist feels it a duty he owes both to his profession and to himself, as an old officer who has devoted his best days to a perilous and laborious line of service, peculiar and distinct in character from that of any other officer of his class, with the single exception of the late Sir John Richardson, who so justly obtained his Inspectorship and "Good Service Pension," on his retirement from the Service.

9. Further, That your Memorialist feels satisfied that such claims as his need only to be respectfully and properly brought under the notice of the Duke of Somerset, to secure for them the reward he asks, as an old officer who has been devoted to the service in which his father served long and well, and finally lost his life in the wreck of his Majesty's ship *Defence*, of seventy-four guns, up the Baltic, of which ship he was Surgeon.

10. Your Memorialist, however desirous of serving his time, has no wish to press for an appointment to the prejudice of any brother-officer, whose claims to it may be presumed to be greater than his own, but to attain his position on the list of Inspectors on Half-pay, which he humbly and most respectfully hopes that his Grace will be pleased to promote him to.

And for which your Memorialist will ever pray.

No. 25.

Letter from PROFESSOR OWEN to GENERAL SABINE, President of the Royal Society.

British Museum, July 18th, 1865.

MY DEAR SABINE,—The new and rare forms of Antarctic mammals and birds shot and preserved by Dr. McCormick, when Surgeon and Naturalist to Sir James Ross's expedition, and which are described in the Zoology of the *Erebus* and *Terror*, as well as in the osteological catalogues of the Museum of the Royal College of Surgeons, long since impressed me very favourably with his attainments and efficiency as a Naval Surgeon. The high state of health of every crew under his sanitary supervision has been favourably reported on in the official documents at the Admiralty; but it is to Mr. McCormick's relations to science, in the discharge of his duties as Surgeon in Polar Expeditions, that I would refer, as mainly moving me to ask you to favourably consider his desire that the very reasonable Memorial he proposes to submit to the First Lord, should have the important advantage of being indorsed or forwarded by you, now the chief, if not sole surviving judge of what the gallant men endured and hazarded in those renowned Expeditions. In hope that you may see reason to give Mr. McCormick this aid,

I remain,

Very faithfully yours,

RICHARD OWEN.

No. 26.

The Royal Society, Burlington House, London, W.,
July 19th, 1865.

MY DEAR SIR,—I am extremely glad to have an opportunity of testifying the high sense which is entertained by the cultivators of Natural History of the Arctic and Antarctic regions, of the value of your scientific services in the several Naval Expeditions which you have accompanied.

I entirely agree with Professor Owen, whose note I return, and than whom there cannot be a better judge, that your long and varied services in both hemispheres entitle you to the *highest* consideration from those who have the privilege of conferring the professional recompense which may be due to the zealous and devoted manner in which you have performed the duties which have been entrusted to you.

I remain, my dear Sir,
With great respect and regard, very sincerely yours,
EDWARD SABINE,
President of the Royal Society.

Dr. McCormick,
Deputy Inspector-General of Hospitals and Fleets.

No. 27.

Admiralty, July 28th, 1865.

SIR,—With reference to the survey held on you on the 14th instant, I am commanded by my Lords Commissioners of the Admiralty to acquaint you that they have been pleased to place you on the Retired List of Deputy Inspectors-General.

I am, Sir,
Your obedient Servant,
R. G. ROMAINE.

No. 28.

Letter to HIS GRACE the DUKE of SOMERSET, K.G., First Lord of the Admiralty.

Junior United Service Club,
July 31st, 1865.

MY LORD DUKE,—I venture respectfully but earnestly, in the painful position to which the summary removal of my name from the "Active" to the "Retired" List of Deputy Inspectors-General has consigned me, without any recognition whatever of past claims, to appeal to your Grace's own sense of justice, feeling assured from the considerate hearing which those claims were favoured with at your Grace's last levée, that this appeal will not be made in vain from an old officer, devoted to the service, who has been arbitrarily cast aside by medical authorities at a survey (in his own profession, too), without so much as any specific assignable cause of inefficiency on his part being given him; and who, still feeling himself as able and willing to serve as ever, reported himself ready to go out to Hong-Kong, on being asked that question by the Director-General.

I cannot but feel, my Lord Duke, under such peculiar circumstances, with a Memorial, endorsed as it has been by such names as Major-General Sabine, the distinguished President of the Royal Society, and Professor Owen, now lying before your Grace, all that is required is, if possible, to obtain for its consideration a leisure moment from the pressure of public duties, which so fully, I am aware, occupy your Grace's and their Lordships' time, unfortunately for me, at this most eventful crisis to myself.

I yet earnestly presume to hope that your Grace, with the Board of Admiralty, will, in their gracious consideration, not permit any ordinary routine obstacles in such a special case as the present to stand in the way of doing a simple act of justice,—by granting me that step in rank on a compulsory removal from the active list, which has of late been so generally conferred upon almost every other grade on retirement.

The want of routine time, no fault of mine, appears to be the only difficulty to be got over; and, setting aside every personal feeling of my own, allow me, my Lord Duke, out of gratitude to them, to claim at your hands some consideration for the names of the highly distinguished and eminent men who have so generously and ably borne honourable testimony to the claims put forward in my Memorial to your Grace, a favourable recognition of which cannot fail to convey a gratifying tribute of respect to the recommendations of these representatives of science, and to the profession to which I belong: that rewards for *special* services are open to the Medical Officer as to other grades in the service.

I have the honour to be, my Lord Duke,

Most respectfully,

Your Grace's humble and obedient Servant,

R. McCormick,

Deputy Inspector-General of Hospitals and Fleets.

No. 29

Admiralty, August 10th, 1865.

Sir,—Having laid before my Lords Commissioners of the

Admiralty your Memorial of the 31st July, praying to be promoted to the rank of Inspector-General of Hospitals and Fleets, I am commanded by their Lordships to express their regret that they cannot comply with your request.

I am, Sir,
Your obedient Servant,
R. G. ROMAINE.

Robert McCormick, Esq.,
Deputy Inspector-General of Hospitals and Fleets,
Junior United Service Club.

No. 30.

TESTIMONIALS

From the Professor of Anatomy, on complying with the Admiralty Circular of 1845, requiring all candidates for promotion to the Inspectorial ranks to go through a course of capital Operations in Surgery, subsequent to the date of the Circular.

Westminster Hospital School of Medicine,
26th August, 1845.

MY DEAR SIR,—It is with no ordinary pleasure I transmit a Certificate for the course of Practical Anatomy and Surgical Operations you have just finished. I can bear testimony to the enthusiasm you displayed in your daily dissections, and to the precision and dexterity you uniformly evinced in performing the capital operations of Surgery. From the knowledge you possess of Surgical Anatomy, and the higher branches of medical and surgical science, you cannot fail to be appreciated as a most valuable member of your profession by all who have the pleasure of your acquaintance.

I have the honour to remain, my dear Sir,
With respect and esteem, yours,
ROBERT HUNTER.

P.S.—Should you deem the above note of any importance, you are at liberty to use it in any manner you may think proper.—R.H.

No. 31.

Certificate from Captain Sir JAMES C. ROSS, R.N.

I HEREBY certify that Mr. Robert McCormick served as Surgeon of H.M.S. *Erebus*, under my command, from the 10th of April, 1839, to the 23rd of September, 1843, during which time he behaved with diligence, attention, and sobriety, and was always obedient to command.

I further certify, that in addition to his duties as chief Medical Officer of the Expedition, which he fulfilled with great ability, he took charge of the Geological and Zoological collections, which through his untiring industry and zeal have proved of much interest and importance in those departments of science.

JAMES C. ROSS, Captain.

December 9th, 1843.

No. 32.

Letter from Captain (Sir EDWARD) PARRY, R.N.

Alderley Park, December 7th, 1827.

DEAR MCCORMICK,—Your letter, announcing the agreeable information of your promotion, has only lately reached me, in consequence of my absence at Liverpool.

I now hasten to offer you my very sincere and hearty congratulations on this event, and I only regret having had no hand in obtaining for you what your abilities and conduct so justly merit.

I should be very glad to avail myself of your offer, if there were any idea of another Expedition being fitted out, which, however, I can assure you is by no means the case.

I hope the delay of my answer will not have interfered with your obtaining other employment.

With every wish for your future welfare,

I remain,

Very faithfully yours,

W. E. PARRY.

No. 33.

Certificate from Captain (Sir EDWARD) PARRY, R.N.

THESE are to certify the principal officers and commissioners of His Majesty's Navy, that Mr. Robert McCormick served as Assistant-Surgeon of H.M.S. *Hecla*, under my command, on a voyage of discovery towards the North Pole, from the 18th day of November, 1826, to the 1st day of November, 1827.

During which time his conduct was at all times such as to merit my highest approbation, not only in the performance of the ordinary duties of his station, but in various other departments, in which his time could be employed with benefit to the peculiar service in which the *Hecla* was engaged.

Given under my hand, on board the *Hecla*, at Deptford,
this 1st day of November, 1827.

W. E. PARRY, Captain.

No. 34.

From Sir ASTLEY COOPER, Bart., to Captain (Sir EDWARD) PARRY, R.N.

DEAR SIR,—Mr. McCormick is a well-informed Surgeon and very enterprising man, well suited to be attached to a dangerous service, and to join one who has immortalized himself by his genius, intrepidity, and his discoveries, like Captain Parry.

Yours very truly,

ASTLEY COOPER.

September 29th, 1826.

SUGGESTIONS LAID BEFORE THE COMMITTEE ON THE NAVAL MEDICAL DEPARTMENT.

1. Medical Officers attached to Scientific Voyages of Discovery to be included in the promotion conferred on all other classes of officers engaged in these Expeditions.

2. Valuable and distinguished services in such Expeditions to be allowed, in special cases, to count as time, in rendering a Medical Officer eligible for promotion to the inspectorial rank.

3. Good Service Pensions to be more liberally bestowed on the Medical Officers of the Navy, and according to relative rank. Pensions to Inspectors-General to be the same as to Flag-Officers; to Deputy Inspectors-General the same as Captains.

4. Medical Officers to be included with other classes of officers in the distribution of the Greenwich Hospital Pensions, according to relative rank.

5. Medical Officers of the Navy to be placed on an equality with other classes of officers in the Navy, and with the Medical Officers in the Army in the distribution of honorary distinctions and rewards of merit.

6. Deputy Inspectors-General to rank on promotion with full Colonels in the Army, and the Senior Captains in the Navy; and, on compulsory retirement from the Service, to be eligible for promotion (without service time) to the next step in rank, as so generally granted to every other class of officers on final retirement.

R. McCormick,
Deputy Inspector-General of Hospitals and Fleets.

November, 1865.

INDEX.

THE END.

GILBERT AND RIVINGTON, LIMITED, ST. JOHN'S SQUARE, LONDON.

Printed in Dunstable, United Kingdom

91222226R00281